Sustainable Civil Infrastructures

Sustainable Infrastructure impacts our well-being and day-to-day lives. The infrastructures we are building today will shape our lives tomorrow. The complex and diverse nature of the impacts due to weather extremes on transportation and civil infrastructures can be seen in our roadways, bridges, and buildings. Extreme summer temperatures, droughts, flash floods, and rising numbers of freeze-thaw cycles pose challenges for civil infrastructure and can endanger public safety. We constantly hear how civil infrastructures need constant attention, preservation, and upgrading. Such improvements and developments would obviously benefit from our desired book series that provide sustainable engineering materials and designs. The economic impact is huge and much research has been conducted worldwide. The future holds many opportunities, not only for researchers in a given country, but also for the worldwide field engineers who apply and implement these technologies. We believe that no approach can succeed if it does not unite the efforts of various engineering disciplines from all over the world under one umbrella to offer a beacon of modern solutions to the global infrastructure. Experts from the various engineering disciplines around the globe will participate in this series, including: Geotechnical, Geological, Geoscience, Petroleum, Structural, Transportation, Bridge, Infrastructure, Energy, Architectural, Chemical and Materials, and other related Engineering disciplines.

More information about this series at http://www.springer.com/series/15140

Hany Shehata · Harry Poulos
Editors

Latest Thoughts on Ground Improvement Techniques

Proceedings of the 2nd GeoMEast International Congress and Exhibition on Sustainable Civil Infrastructures, Egypt 2018 – The Official International Congress of the Soil-Structure Interaction Group in Egypt (SSIGE)

Springer

Editors
Hany Shehata
Soil-Structure Interaction Group
in Egypt (SSIGE)
Cairo, Egypt

Harry Poulos
Coffey Services
Carlingford, NSW, Australia

ISSN 2366-3405 ISSN 2366-3413 (electronic)
Sustainable Civil Infrastructures
ISBN 978-3-030-01916-7 ISBN 978-3-030-01917-4 (eBook)
https://doi.org/10.1007/978-3-030-01917-4

Library of Congress Control Number: 2018957624

This Springer imprint is published by the registered company Springer Nature Switzerland AG
The registered company address is: Gewerbestrasse 11, 6330 Cham, Switzerland

Contents

About the Editors

Shehata is the Founder and CEO of the Soil-Structure Interaction Group in Egypt "SSIGE". He is a Partner and Vice-President of EHE-Consulting Group in the Middle East and Managing Editor of the "Innovative Infrastructure Solutions" journal, published by Springer. He worked in the field of civil engineering early, while studying, with Bechtel Egypt Contracting & PM Company, LLC. His professional experience includes working in culverts, small tunnels, pipe installation, earth reinforcement, soil stabilisation, and small bridges. He also has been involved in teaching, research, and consulting. His areas of specialisation include static and dynamic soil–structure interactions involving buildings, roads, water structures, retaining walls, earth reinforcement, and bridges, as well as different disciplines of project management and contract administration. He is the author of an Arabic practical book titled "Practical Solutions for Different Geotechnical Works: The Practical Engineers' Guidelines". He is currently working on a new book titled "Soil-Foundation-Superstructure Interaction: Structural Integration". He is the contributor of more than 50 publications in national and international conferences and journals. He served as a Co-Chair of the GeoChina 2016 International Conference in Shandong, China. He serves also as a Co-Chair and Secretary General of the GeoMEast 2017 International Conference in Sharm El-Sheikh, Egypt. 2016

Outstanding reviewer of the ASCE as selected by the Editorial Board of International Journal of Geomechanics.

Professor Harry Poulos' AM FAA FTSE Hon FIEAust Dist M ASCE

pioneering work in pile foundation analysis and design has enabled the world's geotechnical specialists to have a greater understanding of the way structures interact with the ground. His research has enabled a more reliable approach to be adopted for pile design, replacing procedures which previously relied purely on experience and empiricism.

He has applied his research to a wide range of major projects, both in Australia and overseas, including buildings, bridges, tunnels, freeways, mines, airports, offshore structures (e.g. oil rigs), and earthquake-related problems. His work includes the Emirates Twin Towers in Dubai, where his analysis and design of the piled raft foundations provided significant cost benefits for the twin towers exceeding 300 metres in height, the Burj Khalifa, now the world's tallest building, where he was the geotechnical peer reviewer, the Docklands project in Melbourne involving design of remedial pile foundations for one of the high rise residential developments, and the construction of a 700-km-long motorway in Greece using his expertise in slope stabilisation and earthquake engineering.

While retaining his professorial position at the University of Sydney, he joined the Coffey Group in 1989, as the Director of Advanced Technology, and became the Chairman of Coffey International Pty Ltd in 1991, a position that he held for 2 years. In the period from 1998 to 2002, he served as the Director of Technical Innovation and General Manager of Technical Development.

He has long been a contributor to the activities of the international geotechnical community. He was also a long-term member of the National Committee of the

Australian Geomechanics Society (AGS) (1980 to 1995) and its Chairman from 1982 to 1984. He was Committee Member of the AGS Sydney Group, 1971–1976, 1979–2002; Vice-chairman, 1974; and Chairman 1980–1981. He was the Australasian Vice-President of the International Society for Soil Mechanics and Foundation Engineering in the period 1989–1994 and an appointed Board Member of the Society from 2001 to 2005 and is currently the Chair of the Membership, Practitioner and Academic Committee of the Society.

He was recognised by his peers for his contributions to Australian Geomechanics by the Sydney Chapter via the institution of the annual Poulos Lecture in 2002.

He is a recipient of many prizes and awards, including Australia's Centenary Medal (2003) for his services to Australian society and science in the field of geotechnical engineering. His overall contribution to the engineering profession has been recognised formally by the award of Member of the Order of Australia (1993), his election as Fellow of the Australian Academy of Science (1988), his Fellowship of the Australian Academy of Technological Sciences and Engineering (1996), his Honorary Fellowship of the Institution of Engineers Australia (1999), the award of the Warren Prize (1972) and Warren Medal (1985) of the Institution of Engineers, Australia, his selection as the 2003 Australian Civil Engineer of the year, and his selection in 2004 as the inaugural Geotechnical Practitioner of the year.

He gave the prestigious Rankine Lecture of the Institution of Civil Engineers (UK) in 1989 and was invited by the American Society of Civil Engineers (ASCE) to deliver the annual Terzaghi Lecture in 2004. He also received from ASCE the 1972 Croes medal, the 1995 State of the Art Award, and the 2007 Middlebrooks Award. In 2010, he was elected as a Distinguished Member of ASCE, the first Australian Civil Engineer to be so recognised.

Ground Improvement as Alternative to Piling – Effective Design Solutions for Heavily Loaded Structures

Wolfgang Sondermann(✉)

Chairman German Geotechnical Society (DGGT), Essen, Germany
w.sondermann@kellerholding.com

Abstract. Well-designed ground improvement options without compromising on stringent design criteria may effectively replace conventional foundation solutions for a wide range of applications involving heavy loads and structures sensitive to settlement. Case studies illustrate the application of different ground improvement methods used in different projects and soil conditions by adopting of advanced design considerations to achieve optimal solutions for the project and customers benefit. Beneficial returns in terms of lowering construction costs and shortening construction time with consistency in quality are discussed where applicable in relation to the required capabilities and experiences of an organization to deliver the alternatives. Opportunity management gives the chance to achieve the project scopes for the client by optimizing cost, time and quality and recommendations are summarized to foster the chance to identify optimal geotechnical solutions.

1 Introduction

More and more civil structures are commissioned as economies around the world grow and the requirements for improved infrastructure accelerate. This inevitably leads to scarcity of land and demands to construct over soft ground, be it formed naturally or artificially. Nowadays, the subject of soft ground engineering attracts a lot of interest and it involves state-of-art engineering practices. Deep foundation using piling techniques is a frequently adopted foundation solution in such site. However, it may not be the only solution and often it is associated with relatively high cost as compared to other alternatives. In some occasions, deep foundation could even be an ineffective solution to adopt in soft ground. For a given scenario, there exists an optimal foundation solution that is technically sound, fit for its purpose and most economical. With rigorous design analysis and careful project execution, ground improvement also in combination with piles would be an efficient and more cost effective alternative to piled foundation only.

H. Shehata and H. Poulos (Eds.): GeoMEast 2018, SUCI, pp. 1–25, 2019.
https://doi.org/10.1007/978-3-030-01917-4_1

2 Best Practice and Evaluation Criteria's for Optimal Solutions

If ground is to be subjected to additional loads which exceed its previous load levels, the geotechnical design requirements have to be established to determine its suitability. Various codes and standards define the limit state conditions influenced by the ground:

- ultimate limit state design (position, design of load bearing members, total load bearing capacity of ground)
- serviceability limit state design (deformation, displacement),
- durability requirements of the products, materials used to construct

which have to be taken considered in the geotechnical design. If it is found, in the course of the geotechnical design, that the above limit state conditions are likely to occur when subjecting the ground to the planned loads, the following measures can be taken:

- unsuitable ground can be bypassed by certain construction measures,
- unsuitable ground can be removed and replaced by suitable ground,
- unsuitable ground can be made suitable using improvement measures.

When selecting the appropriate measures, in addition to taking into account the durability requirements, it is necessary, to determine the soil improvement requirements from the ultimate and serviceability state design in accordance with the geotechnical design categories. The general assessment criteria′s for any ground improvement technology or method are:

- predictability of the technology as a reliable method
- repeatability of the process
- design-ability of the method in combination with the subsoil conditions
- operation as a controllable procedure (documented, monitored)

to evaluate the achievements of the improvement technology as well as the limitations (Kirsch and Sondermann 2003).

3 Ground Improvement Procedures

As a general rule, the desired results of ground improvement are as follows:

- increase in density and shear strength, having a positive effect on all stability problems
- reduction of compressibility, which has a positive effect on deformability
- influencing permeability for the purpose of
- reducing water inflow/water outflow
- increasing deformation speed
- improving homogeneity

Many ground improvement techniques have been developed to treat soil particles to achieve better geotechnical properties. These techniques can be divided into the following categories (Sondermann and Kirsch 2018):

Remove & replace: it is perhaps the oldest and simplest technique. This method involves removing and/or replacing weak soil with better material which has suitable properties for the use. It is generally applied to relatively shallow depth (usually up to 2 m). With more stringent environment protection regulations imposed by authorities, it has become more costly to dispose the weak soil in many countries, especially when the soil is contaminated with hazardous material.

Compaction/densification: this method improves soil properties by rearranging soil particles to denser state using mechanical means. Although surface compaction can be applied to many types of soils, deep compaction can only be effectively and economically used in cohesion-less soils.

Drainage/consolidation: soft soils are generally associated with high water content. Soil strength can be improved by forcing out the water. However, the process often takes a very long time in thick layer of cohesive soil due to low permeability and long drainage paths. To accelerate the process, drainage paths can be introduced. It is almost always used with surcharge, which preloads the soil to a higher degree of consolidation so it can carry intended loadings with little or no further consolidation.

Reinforcement: stiff elements can be introduced into weak soils to carry loads or redistribute loads so less loads are carried by the weak soils. Elements such as geotextile can also introduce tensile capacity into soil mass. Understanding the interactions and load transfer mechanism between stiffer elements and soil mass is vital in design.

Modification: soil properties can be completely changed by introduction of binding agents such as cement or lime. The modified soil would behave similar to very stiff to hard soils or even like low strength rock or concrete, depending on the soil and the amount of binding agent added.

Others: other ground improvement techniques such as ground freezing utilizes frozen water as binding agent to hold soil particles together, while electrical methods like electro-osmotic technique use electrical current to stabilize soil. These methods are less commonly used as compared to the other techniques.

The selection of the most suitable ground improvement method accompanied by the appropriate process sequences for any specific application must always be based on a thorough technical study and economic comparisons and references to the application limitations of the different methods.

4 Geotechnical Opportunity Management

In the context of holistic opportunity management, the project scope has to be achieved taking into account cost, time and quality (Ahmed 2001). In general it is also necessary to weigh the advantages and disadvantages to exploit opportunities in the optimization process and enable the evaluation of potential conflicting tendencies in individual areas (quality optimization requiring costs and time increase versus cost optimization in conjunction with lower quality, etc.). This process is naturally highly dependent on the ability to recognize opportunities and the willingness to exploit them. Last but not least the expertise and experience available in the organization play a decisive role to achieve optimal solutions considering the engineering and operational capabilities of the organisation (human factor).

The contract shall be designed accordingly, so that all participants have a common interest in exploiting these opportunities and share the risks incurred by exploiting them. To manage the site specific opportunities and risks optimally, the mutual interaction and influence of

- site specific characteristics,
- manufacturing technology and process,
- plant and equipment used for construction, and
- personnel (competence and experience)

are of critical importance (Raabe and Wawrzyniak 2015). These four influencing factors define a complex system of mutual dependencies in terms of technical as well as economic aspects and can offer substantial potential for optimization, even during the initial phase of the project. This opportunity can only be exploited when there is adequate geotechnical competence involved in the initial phase of the project (interdisciplinary planning) (Schubert and Bergmair 2002). The greatest opportunities for obtaining an optimal project design are encountered when determining the site characteristics in particular. Different manufacturing processes, even for the same product, have different application limits that often depend on the subsoil characteristics at the site, for example for the selection of a drilling method.

The application of various ground improvement methods and the limits of their application are determined by the site characteristics. In addition to the manufacturing process, also plant and equipment can offer opportunities for an optimized design of a project, for example when using equipment offering optional, alternative methods with fast retrofitting capabilities in case such a risk actually materializes.

To facilitate a meaningful comparison of the technical and economic application of a method, all the necessary basic and boundary conditions relevant to the specific project have to be considered:

- surface area and volume of the ground to be improved
- type of utilization, load application according to type, magnitude and location
- availability of materials, equipment and personnel
- availability of construction personnel with relevant technical know-how of the proposed improvement method
- influences on the environment and the neighborhood
- accessibility, ability to drive on the site in question as a function of the type of utilization,
- environmental protection regulations (ground water, soil, air, noise).

Finally, the competence and experience of the personnel also play a role that should not be underestimated when dealing with the previous three influencing factors and the complex interactions between them. If opportunities are to be exploited, then they need to be recognized as such promptly, which in turn requires depth of experience.

To design an geotechnical opportunity in the interests of all partners involved in addition to the issues mentioned above, the geotechnical risk factors such as:

- (changes in the) soil conditions and properties at the site,
- technical product requirements and design methodologies,
- restrictions in the design of the plant and equipment, and
- environmental factors having an influence

should be addressed, defining the starting point for the optimization process. Before an optimal geotechnical opportunity can be designed, it is essential to determine the site characteristics first (Gudehus 2002). To be able to consider the widest possible range of variants and alternatives for the optimization task, the site investigation should be designed for this purpose (depth of the boreholes, soil sampling method, soil mechanical laboratory tests). Optimal opportunity management that includes the examination of different variants (method/technologies) can naturally encompass a wider range of variants the more precisely the parameters reflect the reality at the site (Witt 2002).

This process requires the study of alternative geotechnical solutions and their requirements on the soil mechanical parameters early on while planning the site investigation, which means that even in this phase, geotechnical expertise can open a range of opportunities. Furthermore, opportunity management is naturally controlled by the requirement of the construction task. In addition to assessing the consequences of damage in connection with the reliability index, it contains recommendations for detailed quality monitoring measures for the planning phase as well as the design phase that should be followed and used as a basis for opportunity management. If the geotechnical properties and requirements on the construction project should be brought into alignment in the next step so that a high quality, economical, and ideally scheduled solution need to be found, then geotechnical methods for solving the problem at hand should be compared to each other and potential alternatives evaluated. During this process the opportunities as well as the risk inventories of individual alternatives also need to be examined and evaluated. The most economically acceptable alternative does not necessarily have to be the one with the lowest risk inventory, and the lowest probability of occurrence may not be associated with the alternative with the lowest influence on cost, schedule and quality. To perform opportunity management in an organized manner in this complex system of geotechnical alternatives, expertise and competence in the following areas should be integrated into the project as early as possible:

- a wide range of geotechnical technologies
- the limits of application of the individual methods (technology, devices, etc.)
- the risk inventory of each method in connection with the ground conditions
- efficiency (cost) of the various methods and applications

If, for example, an excavation pit with foundation piles for the superstructure has already been designed and the superstructure design is also based on this concept, then there will only be a very small opportunity to replace the pile foundation through diaphragm wall elements when awarding the excavation pit contract (Fig. 1).

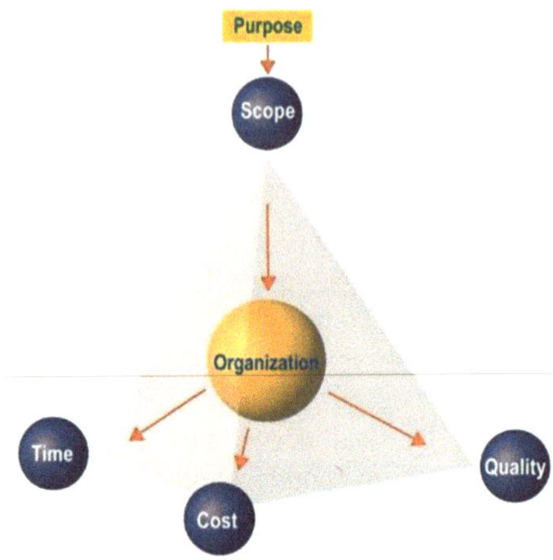

Fig. 1. Factors influencing geotechnical opportunity management

5 Combination of Geotechnical Technologies

In standard design practice, a distinct gap exists between piling and ground improvement (GI) solutions used for foundation support. Piling is typically adopted for heavily loaded structures, such as high-rise buildings, bridges, large commercial centers, silos and other industrial objects, for which stringent bearing capacity, stability and settlement criteria must be satisfied. Conversely, GI solutions are usually considered to be of no or limited use for heavily loaded structures, mainly due to the perceived inability to limit total and differential settlements to an acceptable level. Undoubtedly, the predominant use of piling for foundation of heavily loaded structures is reasonable and has a long tradition. It should be noted, however, there is also a range of applications where piling and GI solutions may actually overlap. This is illustrated in Fig. 2, which shows tentative ranges of GI and piling solutions in a schematic load vs. settlement graph.

For an exemplary GI design, marked as point 1, the associated settlement corresponds to a relatively susceptible foundation support. For the same foundation and loading, a significantly stiffer GI design can be also applied, for example, by increasing the area improvement ratio of the treatment. The resulting reduced settlement is represented by point 2. Between points 1 and 2, other GI solutions exist for the same loading. Taking into account different combinations of foundations, loads and GI methods a range of possible GI solutions can be encircled by connecting a family of corresponding points ‘1’ and ‘2’. A complementary range of piling solutions can be drawn in a similar way. The combined figure obtained reveals an overlapping area where piling and GI solutions may both be applicable for a given range of loads without compromising functional design criteria. In this respect, it should be noted that

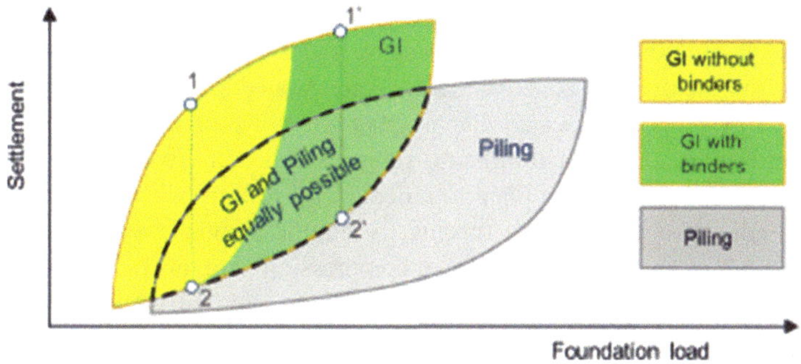

Fig. 2. Tentative ranges of piling and ground improvement solutions for foundation support

GI solutions usually offer shorter execution times and lower construction costs, and are therefore attractive to clients and contractors. In many situations they also reduce the project's carbon footprint.

6 Examples of Geotechnical Opportunity Management

6.1 Sea Tower Gdynia

The underlying projects involve different ground conditions, structures sensitive to total and differential settlements, and heavily loaded foundations in the form of slabs and separated footings. Densification of a loose cohesionless soil deposit by applying VC is one of the most effective GI methods. Figure 3 illustrates the use of VC to support a high-rise building (Sea Towers) in Gdynia, constructed behind a gravity-type quay wall.

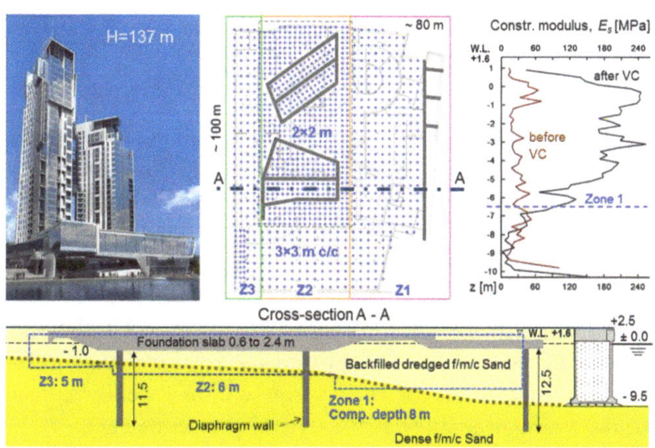

Fig. 3. Compaction of dredged soil with VC to enhance slab-subgrade interaction

The area was a dredged zone of backfilled marine sediments in a loose to medium dense state, composed of fine/medium/coarse sands with some silty inclusions in deepest parts of the dredged zone. While the concentrated loads from both towers were mainly taken by purposely located diaphragm walls, the design aimed to optimize the thickness of the foundation slab utilizing subgrade reaction. To increase ground stiffness below the slab, VC was applied to a depth of 5 to 8 m below the working level, and in two grids of 2 and 3 m c/c spacing. A significant improvement was achieved, allowing the use of an average constrained compression modulus Es of 160 MPa in the design. The GI design was based on FEM 3D soil-structure interaction analysis. There was a good agreement between the observed and predicted settlements (Fig. 4). The VC solution was an alternative to the original design, which comprised 1,312 stabilized columns 0.8 m in diameter and 3.5 m long, on average. The applied scheme was more than three times less expensive than the original solution, and also shortened the construction schedule.

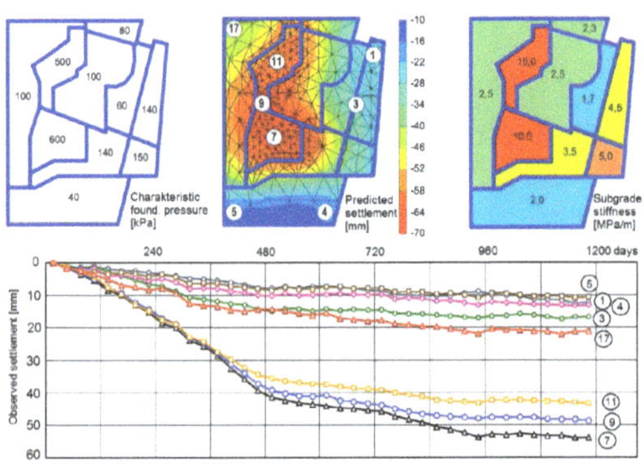

Fig. 4. Predicted and observed settlements, Sea Tower foundation slab

6.2 Iron Ore Storage Facility, Teluk Rubiah, Malaysia

A large international raw material distribution organization was building an iron ore and pellet distribution center on the north-western coast of Malaysia. The project area is located in a wide valley basin, bounded by a range of hills to the north. The distribution center will handle around 90 MTPA (million tons per annum) of iron ore and the total storage space extends about 800 m north to south respectively 1400 m east to west and is about 200 m from the coastline.

The existence of soil with low bearing capability in the area designated for the plant meant that measurements to increase bearing capacity were required in the storage areas in order to reconcile the need for iron ore and pellet storage areas with requirement of the loading facilities as the overall project demand.

As a result of the soil investigation works the subsoil conditions in the stockpiling area was divided into different geotechnical zones. This split into zones A, B and C was based primarily on soil type and layer thickness. Each zone was further divided into subzones (e.g. Zones B1, B2, B3), depending on the variability of the localized soils for the subsequent geotechnical analyses. The entire construction area is underlaid by granite rock formations, which were found at depths of between 3 and 37 m from ground level. This resulted in different treatment zones for the construction site as shown in Fig. 5.

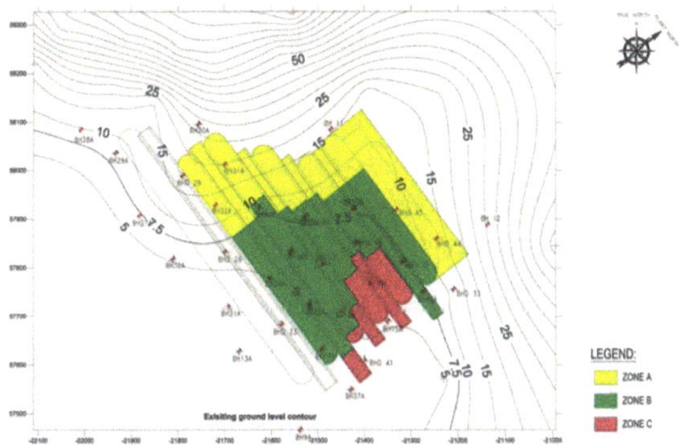

Fig. 5. Geotechnical zones A, B, C and level of rock formation below ground

The loading conditions for the stockyard operations were predetermined for the design of the stockpiling areas using the following framework parameters:

- Bulk density of the iron ore is to be set at 28 kN/m^3
- Average load intensity over the entire stockpiling area approx. 375 kPa
- Peak load in the centre line of the stockpiling area of approx. 532 kPa, corresponds to an overall height of 19 m.
- The maximum load combination (worst case scenario) results from the maximum load of every second stockpiling strip (1, 3, 5, 6).

Long-term performance and settlement criteria were given as requirements for the foundation design and, if necessary, the ground improvement measures in order to achieve the long-term serviceability and stability behavior of the stockpile and loading systems. The criteria which should be fulfilled following the first full load were (Fig. 6):

- to maintain a maximum settlement of 400 mm, indicated in the technical conditions,
- when working out a technical solution for the soil improvement system to respect the max. lateral displacement of 10 mm as tolerance for the rails of the stacker-reclaimer
- and specify in the design the maximum expected horizontal displacement for the piles and the rails along the berms.

As the site conditions are characterized by very difficult geological conditions with very soft soils, along with substantially varying depth stockpiling of heavy iron ore on this kind of soil without additional measures would cause significant displacement (vertical and horizontal). When elaborating the best solution of alternative for the Foundation/Ground Improvement from the several available options for a sustainable reduction in deformation, the following important aspects were considered:

- Technically useful – to transfer both vertical and horizontal loads while adhering to the displacement criteria
- Sustainable – the load-bearing capability does not decrease with time
- Cost effective – should take into account the economic framework;
- Feasible and easy to construct – solution should take into account the reliable technology available in the region;
- Installation procedure and time frameworks – applied technologies should not be burdened with enormous risks or investments;
- Environment – methods should not cause environmental damage to the landscape or bodies of water;

Taking these requirements and limitations into consideration, several different variants were tested:

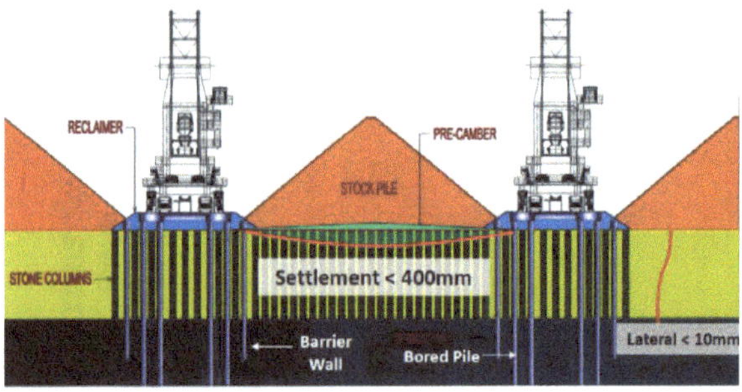

Fig. 6. Deformation criteria for the design of bearing elements

Alternative 1 – vibro stone columns in combination with pre-cast driven piles

This basic concept incorporated the installation of vibro stone columns (Sondermann and Kirsch 2018) in the stockpiling area, over the layer thickness of the soft clay. The stacker-reclaimers were founded on deep pre-cast driven piles. The analysis of this variant showed:

- Vertical movement of 240 mm after initial loading. The ground improvement using vibro stone columns therefore fulfilled the requirement criteria.
- Lateral displacement of 101 mm of the stacker-reclaimer area which exceeded the criterion. In addition, the bending moments of the pre-cast piles exceeded the capacity of the pre-cast piles.

Alternative 2 – vibro stone columns, soil stabilization, pre-fabricated driven piles

Like alternative 1 but using soil stabilization (DSM – deep soil mixing) in the area around the driven piles in order to reinforce the ground around the foundation piles and reduce displacement (Topolnicki 2013). The depth of the DSM treatment varies between 5 m and 10 m. The geotechnical analysis resulted in:

- Vertical movement of 240 mm after initial loading.
- Lateral displacement of 55 mm, which exceeded the criterion. In addition, the bending moments exceeded the acceptance capacity of the pre-cast piles.

Due to the risks associated with this deep mixing technology (Sondermann et al. 2013), this variant was improved by adding a higher replacement relationship of the DSM treatment using grid-adjustments, replacement of the pre-cast driven piles with 1.2 m diameter bored pile groups to support the stacker-reclaimers; and reinforced concrete crossbeams to tie the pile caps together. The analysis of these additions resulted in:

- Vertical movement of 240 mm after initial loading.
- Lateral movement of 10 mm is satisfactory, but bending moments close to the maximum capacity.

Since the DSM technology was judged as locally underdeveloped and adequate equipment was locally not available to the extent needed to cope with the time schedule, another, more conventional alternative was sought.

Alternative 3 – vibro stone columns, bored piles, barrier bored piles

The soil stabilization surrounding the piles of the stacker-reclaimer system was replaced by a barrier wall (bored piles) around the area of the bored foundation piles. The stacker-reclaimer was founded on pairs of piles on each side, both of which were connected by a tie beam. The barrier wall, made of circular, continuous bored piles, was connected by a tie beam (Fig. 7).

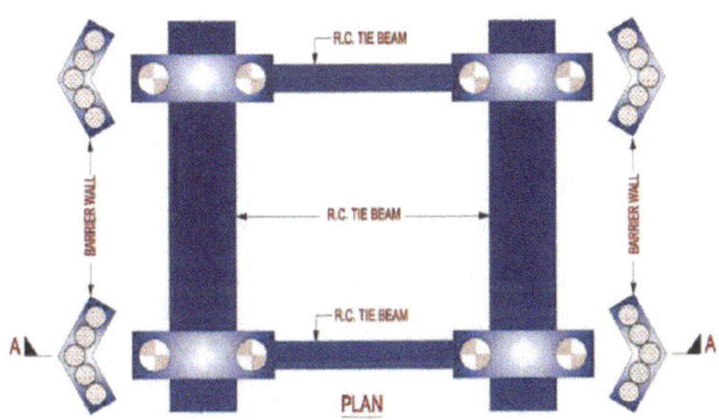

Fig. 7. Alternative 3 with barrier pile walls around foundation plies and capping beams

In order to model the complex geotechnical phenomena in this layout and design, analyses using the finite element method (FEM) were carried out in zones B and C for alternative 3. To minimize the horizontal movement of the loading system, a certain construction/loading procedure was established which was also simulated with the FEM analysis. In this context, the extended elasto-plastic hardening model for the simulation of the soils in the current FEM analysis was used to illustrate the complex behavior of the plastic shearing and isotropic-plastic load process. For the subsequent asymmetrical loads, elastic soil behavior was expected of the zone with higher soil stiffness in the unloading-reloading area (Schanz and Vermeer 1999). The results of the investigations for symmetrical and asymmetrical load conditions show that the stiffness of the barrier construction has a significant impact on the forces and the deformations of the foundation piles as well as on the barrier piles themselves.

In order to verify these impacts and to examine the accuracy and reliability of the 2D FEM analysis results, a complex 3D FEM analysis for zone C was carried out, using the most unfavorable soil conditions of soft clay with loose sand up to approx. 20 m under the ground surface. The soil parameters and simulation of construction sequences remain the same as for the 2D FEM analysis. Figure 8 shows the 3D FEM mesh in the event of maximum asymmetrical loading: stockpiling up to 19 m. In contrast, Fig. 12 hides some soil elements in order to illustrate the spatial configuration of the front barrier wall and the bored piles connected with a 1.5 m × 1.5 m tie beam.

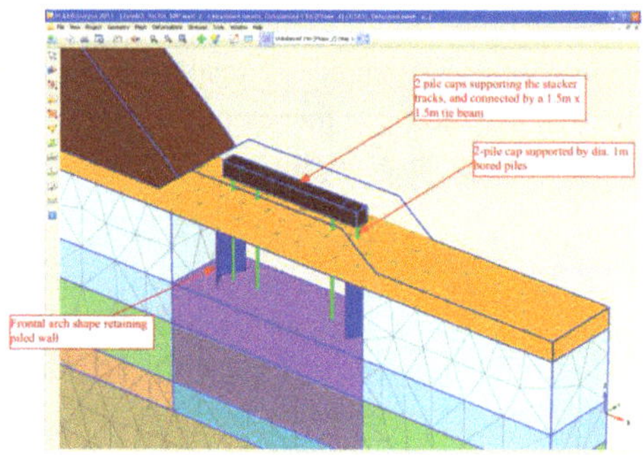

Fig. 8. Spatial configuration of the barrier wall and bored piles

The impact of stiffness and layout of the barrier wall was investigated and showed that too high stiffness in this system is leading to an increase in the diameter of the piles as well as in the amount of reinforcement. The comparison with the 2D FEM results of approx. 9.6 mm verified the accuracy and reliability of the 2D FEM and showed the effectiveness and feasibility of the proposed alternative 3.

2D and 3D FEM analyses were also carried out for zones B and C and the results of the 2D FEM analysis were confirmed by the 3D FEM analyses for a critical section in zone C, which showed easily comparable results. Essentially, the efficiency of the suggested variant depends mainly on the form and stiffness of the barrier construction, which produces a supporting arch effect, in order to achieve an effective minimization of the soil movement and to minimize the effects on the horizontal load of the bored piles. Varying the diameters of the barrier piles and the depth produced an optimal load on the bored piles for the crane track. These effects were also observed by Qiu and Reimann (2013) in very small scale tests, where creating a bearing arch between the barrier piles diverts the horizontal load into these piles.

Because the loading sequence during the stockpiling of the iron ore material on the improved ground was crucial for successful compliance with the deformation criteria, an initial load scheme was elaborated following these design considerations. The instrumentation as an important component of the implementation of the project is based on the following principles:

- The instrumentation should allow the adoption of design and dimensioning measures to be inspected in-situ
- The monitoring program should cover the entire stockpiling area, in order to examine the reaction of the improved subsoil under load and to take any appropriate, necessary measures
- The monitoring data are noted and automatically stored on a server in order to be able to evaluate the data at any time, depending on operational needs
- Communication with the operating team, if the threshold values are reached or exceeded

The information about the soil reactions during the iron ore stockpiling process should be made available for the optimization and management of this process. The instrumentation, with monitoring program, was installed before the first loading and was subject to regular evaluations for subsequent loading cycles for up to 2 years after the initial loading.

6.3 Foundation for Hull Block Assembly Factory, Singapore

A very large single-storey steel frame structure (670 m × 185 m, up to 50 m in height) is being constructed with the structure consisting of storage area with loads up to 160 kPa and factory area that houses heavy overhead cranes and laser cutting and welding equipment, which are very sensitive to differential settlement. The site area was reclaimed about 2–3 years prior to the foundation works. Preliminary soil investigation indicates that the site is underlain by up to 27 m of reclamation fill, mostly consisting of relatively clean sand, overlying original seabed. Up to 15 m of sandy clay was found within the reclamation fill in some area. Underneath the reclamation fill, up to 11 m of soft to firm marine clay was discovered at a few soil investigation locations. Extensive soil investigation was carried out once the decision of adopting ground improvement was made. The main purposes of the investigation are to obtain more information on the sub-soil condition and to identify the extent of weak soil for an optimal ground improvement design. It is revealed that a "valley" about 50 m wide and up to 11 m deep filled with marine clay runs though the foot print below reclamation fill (Fig. 9). A cone penetration test (CPT) result within the area is shown in Fig. 10. It is worthy to note that the marine clay layer only underlies area with relatively small load (20–50 kPa live load) as compared to the storage area (160 kPa dead load). Addition investigations disclosed the soil is under-consolidated and will continue to undergo consolidation process which result in significant settlement even if no additional load is placed.

The structure being constructed is a steel frame structure with cladding roof and façade. The structural columns are arranged in strips with about 9 m spacing between columns and about 20–30 m between strips. These columns support the main structure as well as crane rails that span between columns along columns strips. The columns are founded on reinforced concrete ground beams of over 1 m thick and up to 5.8 m wide.

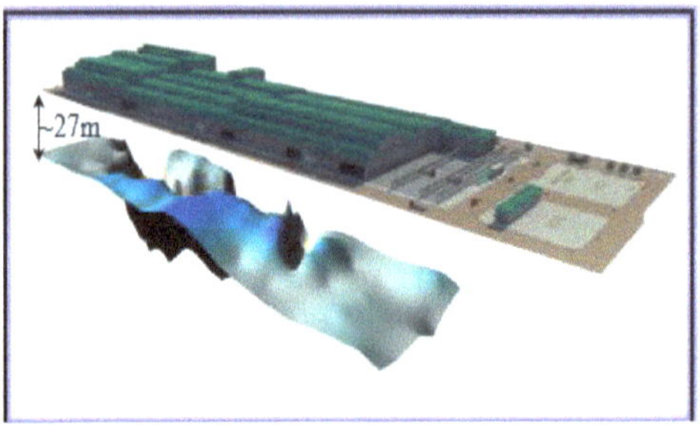

Fig. 9. Underlying marine clay layer

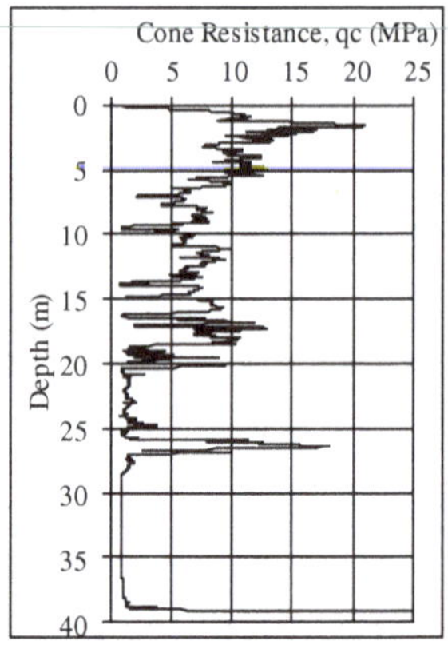

Fig. 10. Typical cone penetration test result

Steel cutting, forming and welding equipment are installed on the factory floor, which are constructed of reinforced concrete or fibre reinforced concrete with localised enhanced reinforcement for heavy machines. The layout plan of foundation beams and slabs is shown in Fig. 11.

The factory building is divided into four parts, which are separated by expansion joints, in accordance to its intended usage: Part 1, Part 2 & 3 and Part 4 are the main factory floor area where steel cutting, forming and wielding activities will be carried out; Part 5 mainly consists of storage area where up to 2 m of steel plate will be stored. The foundation beams and slabs for each part are constructed as one large raft to minimise the differential settlement. Transition slabs of 1 m thickness are also constructed underneath expansion joints to limit the differential settlement.

Foundation Solution

Initially driven pile was considered for supporting the column strips and the slabs would be founded directly on vibro compacted ground. After reviewing the sub-soil conditions, it was determined that this solution was unfeasible due to the presence of the marine clay (Raju et al. 2005). The ground would continue to settle due to

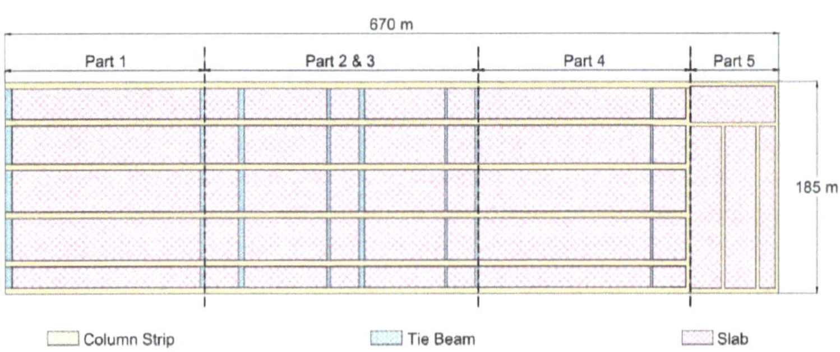

Fig. 11. Foundation raft layout.

consolidation of the clay if it is left untreated. Significant differential settlement would occur between slabs and column strips. To cater this, both slabs and columns would have to be supported by piles, or large bored piles need to be used with slabs being suspended between column strips. Furthermore, due to the presence of thick sand layer above the consolidating clay layer, the piles would have to be designed to take considerable negative skin friction (NSF), which could be greater than the structural capacity of available driven piles. In addition, piles are likely to bear more NSF if the sand is compacted, as more friction would be generated in sand layer. Therefore, piles are not effective for use under such circumstance.

Thick layer of well compacted sand on the top would provide sufficient bearing capacity and a stable platform for the structure, and pre-consolidated marine clay would not settle excessively. Accordingly, deep compaction using vibro technique is adopted to compact the sand layer and prefabricated vertical drain (PVD) with surcharge is used to pre-consolidate the marine clay. The ground improvement scheme is illustrated in Fig. 12.

Instrumentation program is also implemented during the ground improvement work to monitor the consolidation process of marine clay. Since the marine clay is at more than 27 m below ground surface, deep settlement gauges (extensometer with magnetic spiders) are installed to measure the settlement within marine clay layer. The predicted and measured settlement in marine clay layer during surcharge period is shown in Fig. 13. In order to ensure the required degree of consolidation is achieved at the end of surcharge, part of the surcharge was removed and the settlement is observed for about a week.

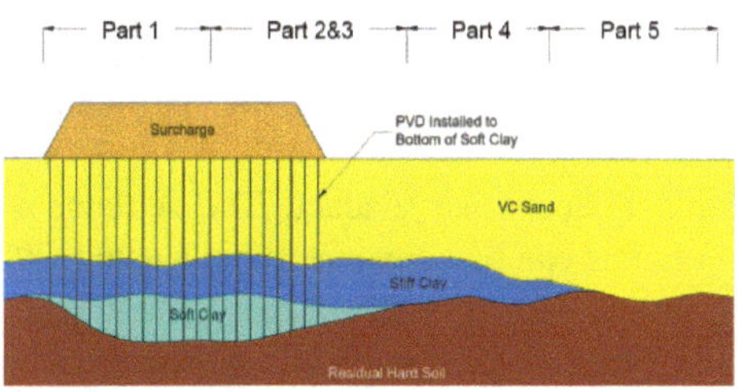

Fig. 12. Ground improvement scheme

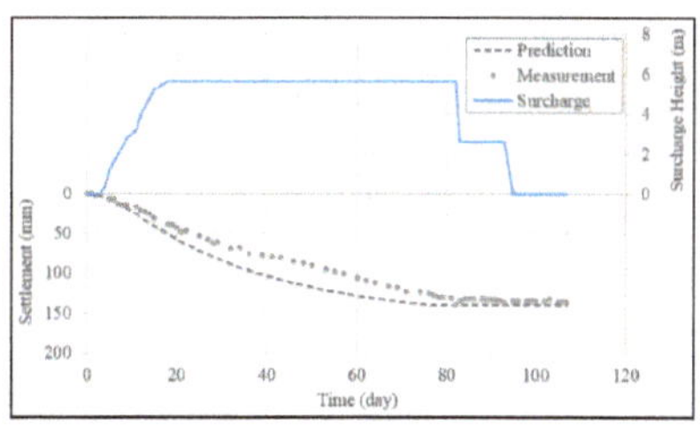

Fig. 13. Settlement - surcharge period

Presence of a narrow strip of thin marine clay layer with about 20 m width and 3–5 m depth was also found below 26 m of sand. 3-D finite element analysis was performed to evaluate whether the clay layer need to be treated. The settlement of marine clay is relatively larger than surround soils at 26 m below ground level. However, similar settlement contour is not reflected on the ground surface (Fig. 13). Instead, the surface settlement corresponds more to the loading pattern rather than the shape of marine clay area. This indicates that the load from ground surface has been arched/bridged over by the stiffer vibro-compacted sand fill above the soft clay. Therefore, no treatment was applied to improve the soft clay (Fig. 14).

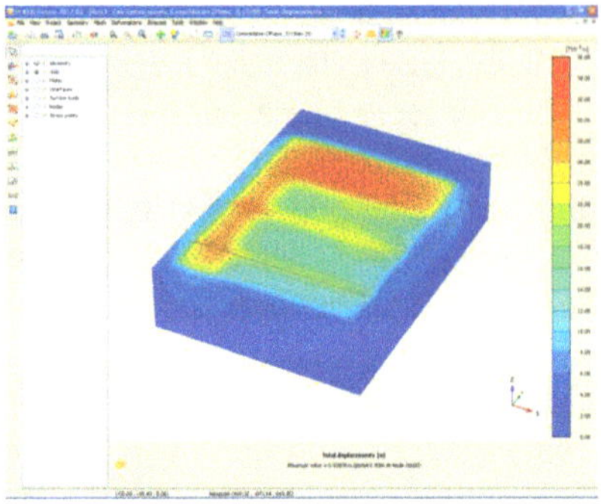

Fig. 14. Long-term ground settlement contour with max. of about 36 mm

6.4 Deep Soil Mixing as Foundation of Road and Railway Viaducts

Since 2002 more than 250 road and railway viaducts have been successfully founded on treated ground by deep soil mixing with the benefit of shorter construction times and lower costs of ground improvement with DSM contributed to substantial economical savings as compared to traditional piling technologies (Fig. 15).

The selection of specific installation patterns of DSM columns requires careful attention to viaduct abutments and foundations loaded with large horizontal forces and overturning moments. Columns installed to improve the stability of foundations should be placed in single or double rows, purposely arranged with regard to critical loading schemes. This increases the system efficacy when compared to single columns, in that the negative effect of local column weakness and the risk of bending failure of the columns are reduced. Overlapping of columns in the individual rows increases bending resistance. Overturning stability can be improved through increasing the number of columns and increasing the length of rows.

Figure 16 illustrates DSM works for a typical road viaduct. The settlement of viaduct foundations supported on DSM columns can be conveniently evaluated with a simplified method, similar to that used for pile groups. For this purpose, the maximum characteristic vertical load applied on the base of the foundation is converted to a uniform pressure, and the foundation is assumed to be rigid. The total settlement is calculated as the sum of: (1) the compression of the deep mixed zone below the foundation, (2) the compression of the underlying ground, and (3) the punching of columns into the underlying soil layer at toe level.

The compression of the deep mixed zone is calculated based on the assumption of equal strains in the columns and the adjacent untreated soil within the foundation area. The compression of soil below the columns is evaluated with a load-spread method using stresses determined in the elastic isotropic half-space below a fictitious foundation assumed at columns' toe level.

Punching is especially important for columns with shallow embedment depth and for ground improvement applications with floating columns. The settlement caused by punching can only be roughly estimated when restricting calculations to sole elastic deformations. For this purpose the column base area is replaced by an equivalent

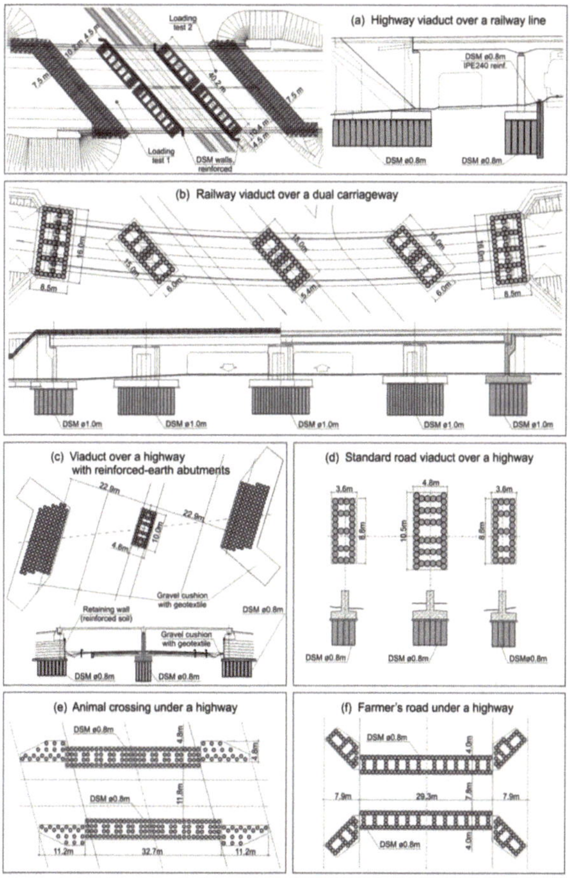

Fig. 15. Installation patterns of DSM columns under foundations of viaducts

Fig. 16. DSM works/exposed columns before placement of a lean concrete layer

square plate, which is loaded with a uniform pressure corresponding to the difference between the stress in the column at toe level and the stress acting at the base of the fictitious foundation.

Since localised plastic deformations are not taken into account in this approach, the resulting settlement due to column base punching is usually smaller than one calculated with elastic-plastic models. For viaduct foundations with complex column patterns and/or loading schemes, consideration should be given to supplementing simplified settlement analyses with finite element (FE) computations.

6.5 DSM as Foundation of Wind Turbines

When soil mixing is performed to support wind turbine foundations the external loads can be transferred down to the bearing layer, resulting in a fixed type improvement, but can be also partly or wholly transferred to the foundation soil when a more interactive or even a floating type of foundation system is desired. The choice of the load transfer system adopted and of the required strength of cemented soil is dictated by functional requirements and associated design criteria.

The geotechnical design of DSM is aimed to determine the final installation pattern and dimensions of improved ground on the basis of relevant Ultimate Limit States (ULS) and Serviceability Limit State (SLS) analyses to satisfy functional requirements of wind turbine manufacturers. It must be noted that the behaviour of wind turbine foundations is generally governed by a high overturning moment. Consequently, foundation tilt is of major concern as well as the overall stability of the wind turbine, which strongly depends on the position of the centre of rotation.

Depending primarily on the adopted arrangement of DSM columns and on the selected design UCS of stabilised soil, which in general may represent hard to semi-hard material, the improved ground is usually considered as a geo-composite system. Deformation and stability analyses for composite ground are often complex and involve non-linear and stress/strain dependent interaction behaviour. Therefore simplified calculations based on semi analytical methods often need to be cross-checked or supplemented by advanced 3D FE analysis to determine more realistic foundation displacements and stresses acting in individual columns.

In practice, mainly blade-based mixing tools diameter 0.6 to 1.6 m and single-shaft rigs have been used to install DSM columns purposely located along the periphery of wind turbine foundation. Both, unreinforced and reinforced DSM columns have been used, as shown in Fig. 17.

In this case, comprising 98 m high wind turbine with rated power of 2 MW, the adopted solution involved 50 unreinforced DSM columns diameter 1.0 m, arranged in two rows along the periphery of a foundation base having a diameter of 17.5 m. To avoid stress concentrations on the trimmed heads of DSM columns, a 50 cm thick sand/gravel cushion has been designed, installed and compacted before application of the lean concrete layer, 10 cm thick. The adopted characteristic compressive strength of soil-mix was 2.3 MPa after 56 days of curing.

The second example, presented in Fig. 18, illustrates the application of 38 no. unreinforced DSM columns diameter 1.6 m, applied for ground improvement below a circular foundation of 20.5 m diameter, supporting 119 m high wind turbine. The

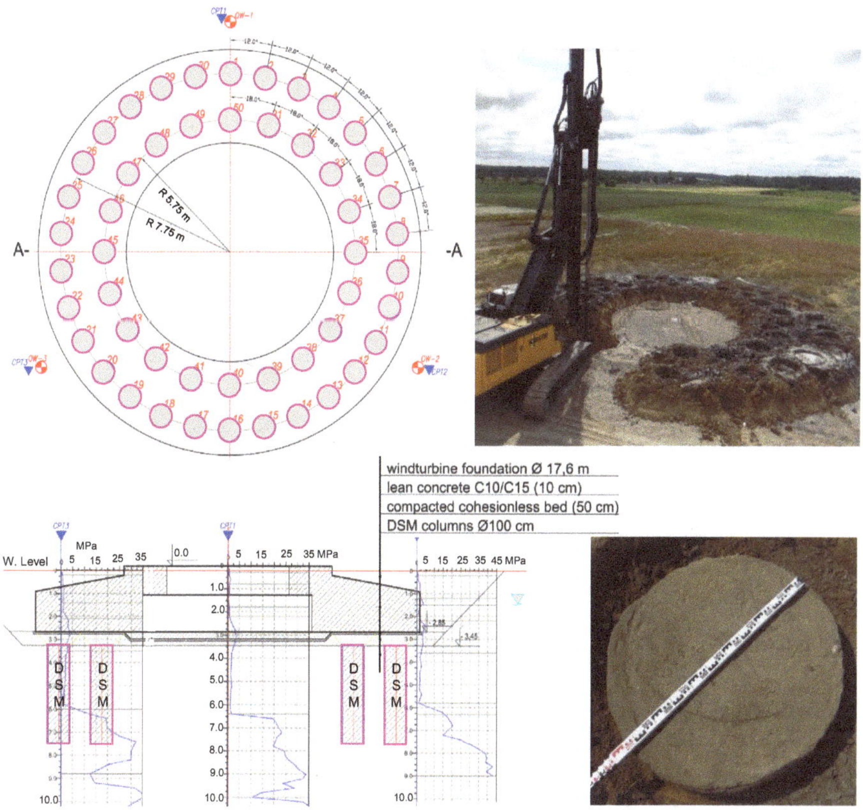

Fig. 17. Wind turbine foundation supported on 50 unreinforced DSM columns diameter 1.0 m (turbine height 98 m)

resulting area improvement ratio is thus 23.1%. The transition layer above trimmed DSM columns, also 50 cm thick as in the preceding case, has been constructed using compacted aggregate or cement-stabilised soil. The adopted characteristic compressive strength of soil-mix was 2.2 MPa after 56 days of curing.

When a high overturning resistance of wind turbine foundation is required, the DSM columns may be reinforced with steel beams and used in a similar way as compression and tension piles, as illustrated in Fig. 19. The adopted solution comprised 40 reinforced DSM columns diameter 1.0 m, positioned in two rows along the periphery of a 16-sided polygon foundation with an average diameter of only 15 m.

Soil mixing was conducted from the bottom of shallow excavations, about 2 m deep. The columns, which were 6 to 8.7 m long depending on turbine location, were designed to take maximum compression and tension forces resulting from the adopted loading conditions. To accommodate for tension forces steel soldier profiles IPE 220, with lengths of 6 to 8 m, were centrally installed in each column just after completion of the soil mixing operation. They extended a minimum of 0.7 m above the final level of column head, allowing for 0.5 m embedment length in the concrete foundation.

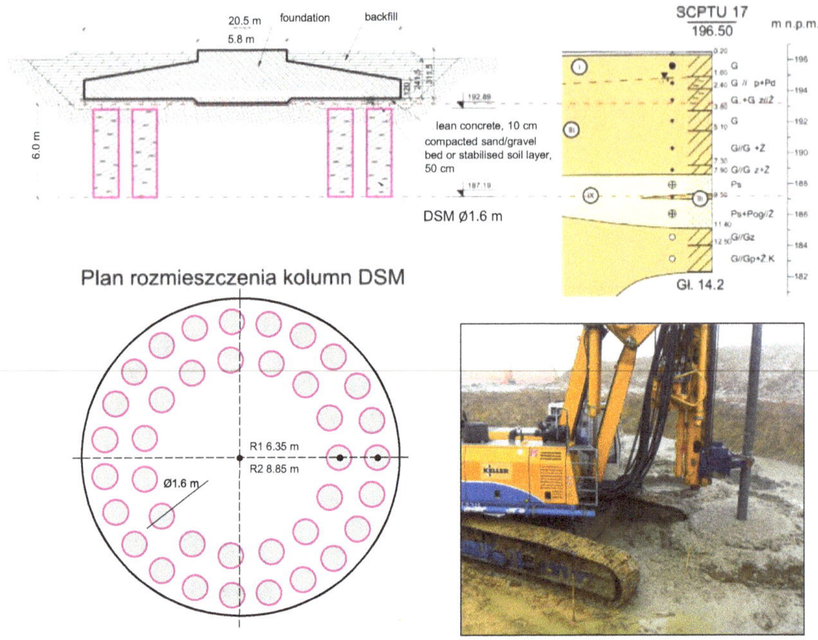

Fig. 18. Wind turbine foundation supported on 38 unreinforced DSM columns diameter 1.6 m (turbine height 119 m)

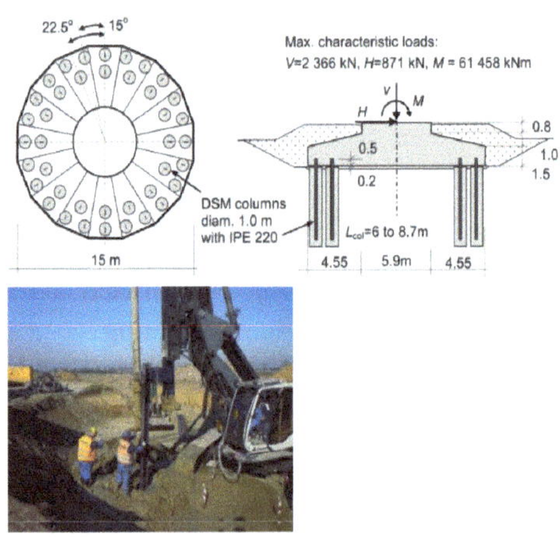

Fig. 19. Wind turbine foundation supported on 40 reinforced DSM columns diameter 1.0 m (turbine height 78 m)

For reinforced columns bond stress of soil-mix must be checked to assure safe transfer of tensile loads to steel solider elements. Taking into account the requirements resulting from compression and tension forces acting on DSM columns, the characteristic compressive strength of soil-mix material in this project was set to 2 MPa after 56 days of curing.

7 Conclusions

The interaction between the ground conditions and characteristics of the various homogeneous areas and technical project requirements determine the portfolio of geotechnical solutions. The determination of realistic parameters has a decisive influence in this case on the selection of potential geotechnical methods and the possible designs. The chances for obtaining an optimal geotechnical design for a project increase disproportionately as the density of the geotechnical information obtained increases. The significant challenges for the future can be summarized in the following five theses:

- **Construction projects are becoming more and more complex:**

In the future, the increasing complexity of projects will probably lead to significantly more attention being paid to experience in such tasks as planning, designing, calculation and execution. Only by integrating this experience into the construction process this complexity will be manageable.

- **Economic optimization requirements will promote global co-operation and increasing global competition:**

In the future, optimizations aimed at economic improvement will also create substantial cost pressure and will force global co-operation. These developments will lead inevitably to increasing international competition and likewise, this increasing internationalization of the market conditions will bring intensified contractual conditions and clauses with it. In order to meet this challenge, improved knowledge of contract and risk management, right from the project management level, will be essential, alongside the increasing requirement for technical capabilities.

- **Increasing demand for automation and the automatic monitoring of building processes:**

With increasing complexity and increasing pressure on efficiency, automation and the automatic monitoring of building processes will become increasingly important in order to maintain the require standard of quality. Minimizing risks of faults and bridging any gaps in experience as necessary, in order to avoid negative consequences, will be a significant task for the future. In the future, automation will create new requirements, on device manufacturers and suppliers in particular, to directly and promptly optimize productivity and quality using improved, process-oriented monitoring in the construction process.

- **Clients and site supervisors will expect a more efficient documentation of the processes and quality:**

Increasing requirements of the documentation of construction processes will lead inevitably to improvements in existing quality and risk management systems and be developed parallel to the requirements for automation. An even closer cooperation between users and manufacturers of construction equipment should in most cases be adapted to these future tasks for process cycle documentation using measuring and sensor technology.

- **Environmentally-friendly technologies are becoming more attractive:**

The development and advancement of resource-saving technologies and building processes will become a significant challenge for the entire building industry, but particularly for geo-technology. Alongside the reduction in the consumption of materials and of emissions (noise, dust, vibration…), the reduction of CO_2 emissions and also the improvement of the logistics process will be at the forefront.

Innovative solutions in use of new or improved technologies as well as the changing and adjusting of existing technologies to suit the specific requirements of the project can contribute to generating an increase in value for the owner. These approaches can also play their part in meeting the future challenges to reduce complexity, to increase quality through automation and documentation as well as to strengthen the global competitiveness.

8 Recommendations

Considering the future challenges as general requirements for changes innovative solutions could show a way out of this problematic issues. But innovative alternative solutions require a more precise management of alternative opportunities. Optimal opportunity management largely depends on the information density and the expertise of the project members and also has a permanent effect on the success of the project. Opportunity management gives the chance to achieve the project scopes for the client by optimizing cost, time and quality. The following recommendations should foster the chance to identify optimal geotechnical solutions:

- **Early high quality and extensive investigation on site**

A good and comprehensive investigation of the soil at the site can point out opportunities for the use of alternative processes and methods regardless of the geotechnical problem at hand. Consolidating the knowledge gained later on often fails due to limited time. The supposedly minor economic disadvantages of performing a comprehensive soil investigation at the site are compensated for over the course of the project in almost all cases through optimized opportunity management.

- **Design of the alternatives using high quality methods**

A detailed examination of alternative geotechnical solutions using higher quality design methods and stress strain laws generally leads to savings during construction in time as well as in costs and improve quality in addition.

- **Detailed planning as part of the planning process**

The purpose of creating a project plan ready for implementation is to ensure all conditions relating to the feasibility of construction (accessibility, equipment and plant, material logistics, environmental influences and degradation, etc.) are checked in detail, and thus ensure the risks resulting from them are minimized, in order to optimally take advantage of the opportunities available.

- **Consideration of experience and expertise of people involved in the project**

Optimal opportunity management thrives off the opportunities that are recognized as early as possible in the project life cycle (van Staveren 2013), which depends substantially on the level of expertise and experience of the engineers involved and available. For this reason, soft skills should not be ignored when selecting the participants in the project.

- **Early specification of the optimization objectives**

Clearly defined priorities during the decision-making process have a major influence on the ability to exploit opportunities, and changing optimization objectives or priorities as the project progresses generally lead to substantial disruptions and higher project risk.

- **Transparent project information system including risk and opportunity registers**

A project information system can substantially minimize the risks resulting from misinformation, especially when the tasks are complex. A transparent information system accessible to all participants, including a risk and opportunity register makes a permanent contribution to the success of an opportunity management system.

For geotechnical engineers the main future challenges are to be familiar with theoretical as well as practical applications of advanced technologies to identify for a specific project the optimal combination of technologies including the best design approach to stay competitive.

To cope with this more general and conceptual formulation of the task the below summarized **capabilities and competences of geotechnical engineers** to create optimal geotechnical solutions are required:

Technical skills:

- deep knowledge of technologies and processes including the limits of application of the different technics
- excellent judgement on subsoil characteristics and stress strain relationships of soil materials

- advanced design capabilities including judgement regarding application of design model and approach
- understanding the execution process and the impact of the process on adjacent structures and the environment
- fair judgement on technical requirements versus products involved including alternatives

To cope with these requirements in addition to the technical skills geotechnical engineers should show capabilities in:

Soft skills:

- Planning and organizing
- Analysing and structuring
- Accuracy and reliability
- Team work and relationship with colleagues

to identify optimal solutions with the teams from different disciplines involved in a project. Those geotechnical engineers developing a sound and **perfect combination of the technical expertise and the soft competences** will be fit for future challenges. As a geotechnical engineer is should be the goal not only to sharpen the geotechnical knowledge but also acquire and shape interpersonal skills.

References

Ahmed, S.M.: Measurement of Construction Processes for Continuous Improvement. State of Florida, Department of Community Affairs (2001)

Gudehus, G.: Geotechnisches Risikomanagement aus bodenmechanischer Sicht (Geotechnical risk management from a soil mechanics perspective). In: Proceedings of the 12th Danube-European Conference in Passau, pp. 21–29 (2002)

Raabe, E.-W., Wawrzyniak, C.: Chancen und Risiken der Vertragsgestaltung für geotechnische Leistungen, (Opportunities and risks of contract designs for geotechnical services. Bautechnik **92**(2015), 624–630 (2015)

Schubert, P., Bergmair, M.: Geotechnische Risikobewertung und Risikomanagement - Methodologie und Fallbeispiele (Geotechnical risk evaluation and risk management - methodology and case histories), Felsbau 20, no. 5, p. 31ff (2002)

van Staveren, M.: International State of the Art Report on Integration of Geotechnical Risk Management and Project Risk Management, Part 1-Report, ISSMGE TC 304 - Task Force 3 (2013)

Witt K.-J.: Das Baugrundrisiko aus geotechnischer und vertragsrechtlicher Sicht (Foundation risks from a geotechnical and contractual perspective). "Damage Cases" Seminar of the German Association of Road and Transport Engineers (VSVI), November 2002

Topolnicki, M.: In-situ soil mixing. In: Kirsch, K., Bell, A. (eds.) Ground Improvement, 3rd edn, pp. 329–434. CRC Press, Boca Raton (2013)

Qiu, G., Reimann, K., Grabe, J.: Zur Reduktion des aktiven Erddrucks infolge Verdübelung mittels Pfählen, Pfahl-Symposium Braunschweig 2013, report from the Institut für Grundbau und Bodenmechanik Technische Universität Braunschweig (Institute for Soil Mechanics and Geotechnical Engineering Technical University of Brunswick), vol. 96, pp. 327–337 (2013)

Raju, V.R., Sondermann, W.: Ground improvement using vibro techniques in ground improvement case histories. In: Indraratna, B., Chu, J. (eds.), pp. 601–638 (2005)

Schanz, T., Vermeer, P.A.: The hardening soil model - formulation and verification. In: Brinkgreve, R. (ed.) Beyond 2000 Computational Geotechniques, pp. 281–296. Balkema, Rotterdam (1999)

Sondermann, W., Kirsch, F.: Baugrundverbesserung in Grundbautaschenbuch, 8. Auflage, Teil 2. In: Witt, K.-J. (ed.) Ernst & Sohn, Berlin (2018)

Sondermann, W. et al.: Design risks of ground improvement methods including rigid inclusions. In: Proceedings 18th South East Asian Geotechnical Conference, Singapore, 29–30 May 2013

Behavior of Ordinary and Encased Stone Columns End-Bearing and Floating in Soft Clay (Numerical Model)

Mahmoud Ali(✉)

Construction Research Institute, Cairo, Egypt
eng_mahmoud1991@yahoo.com

Abstract. Stone column is used for strengthening the soil foundation under the raft. Stone column in soft clay that have undrained shear strength less than 15 kPa is not efficient due to weak lateral confinement. The encasement is used to reduce lateral confinement. The soil may be extend to 40 m such as in Port Said region so the partial penetrating column is a suitable solution of this case. In this Paper, finite element analyses were conducted with using the software PLAXIS 3D. This paper discuss column length effect on vertical settlement, excess pore pressure, and stress concentration ratio with respect to encasement.

Keywords: Encased sand column · Geotextile · Soil reinforced End bearing · Soft clay · Rigid footing · Plaxis 3D

1 Introduction

Water storage tank rested on deep soft clay may cause large settlement. A lot of soil improvement techniques developed to minimize the settlement and increase soil foundation bearing capacity. One of the most efficient techniques is stone column. The main target of using the stone column technique is to reduce settlement, increase bearing capacity, and accelerate the consolidation process.

The stone column is not efficient in soft clay having undrained shear strength less than 15 kPa because the weak soil don't provide sufficient lateral pressure so the lateral deformation increase. The encasement of the stone column with geosynthetic material provide lateral confinement. Experimental studies performed on un-encased granular columns (Juran and Guermazi 1988; Gniel and Bouazza 2009) demonstrated that the stress concentration varied mainly between 2 to 3 but may increase up to 20 in some cases of geosynthetic-encased granular columns.

Malarvizhi and Ilamparuthi (2008) reported that bearing capacity of encased stone column was 1.5 to 2 times that of ordinary stone columns. Murugesan and Rajagopal (2006) reported that the encasement length larger than twice the diameter of the columns did not develop more improvement. Gniel and Bouazza (2009) Concluded that individual columns failed by radial deformation beyond the level of encasement where as in partially encased group of column, bulging occurred along the full length of non-encased section.

H. Shehata and H. Poulos (Eds.): GeoMEast 2018, SUCI, pp. 26–38, 2019.
https://doi.org/10.1007/978-3-030-01917-4_2

This study investigates the performance of stone column under storage tank and also discuss the effect of column length and column encasement in terms of vertical settlement, excess pore pressure, and stress concentration ratio.

2 Finite Element Model

3D finite element model was conducted to understand the behavior of stone column under water storage tank raft. The commercial FE package PLAXIS 3D was used for modelling. The 3D model shown in Fig. 1

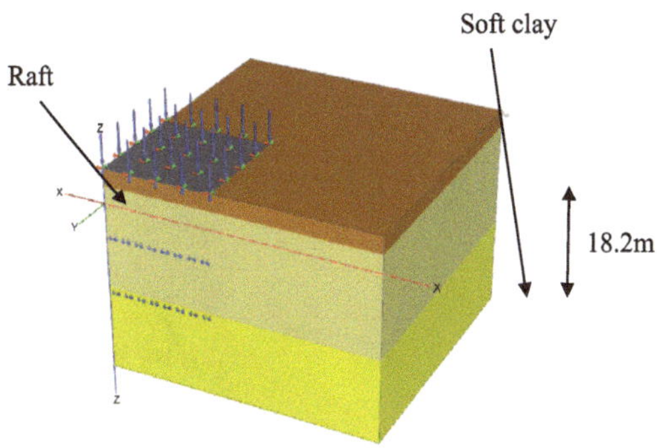

Fig. 1. 3D finite element model of raft rested on stone column

The footing with dimensions 40 m long × 40 m wide rested on stone column with columns spacing equal 2.5 m with rectangular arrangement pattern. The column diameter was equal 1 m, as shown as Fig. 2.

As displacement boundary is concerns, no displacements in the directions perpendicular to the symmetry planes and to the base were allowed. For the hydraulic boundary condition, the phreatic level was set at the top surface of the soft clay layer to generate a hydrostatic pore water pressure profile in the domain. A zero pore pressure boundary condition was applied at the top. The left boundary was assumed impervious to consider the fact that no flow entered or left the symmetry plane. Since the right boundary was too far from the footing to have significant influence on the results, it was set impervious. The finite element mesh was developed using 10 node tetrahedral elements to represent soft clay, stone column, and footing. The geosynthetic reinforcement was modeled as geogrid element available in PLAXIS 3D, composed of 6 node triangular surface elements. The finite element distribution of the mesh was set to "fine" with local refinement close to the stone columns. The geosynthetic was modeled as geogrid element available in PLAXIS 3D having axial stiffness only. After generation of initial stress and pore water pressure, stone column was model by replacing soft soil element by stone column, and the geosynthetic reinforcement was added as wished in place. The raft was left 3months to consolidate after tank construction. The water was filled in tank in 10 m water height increment.

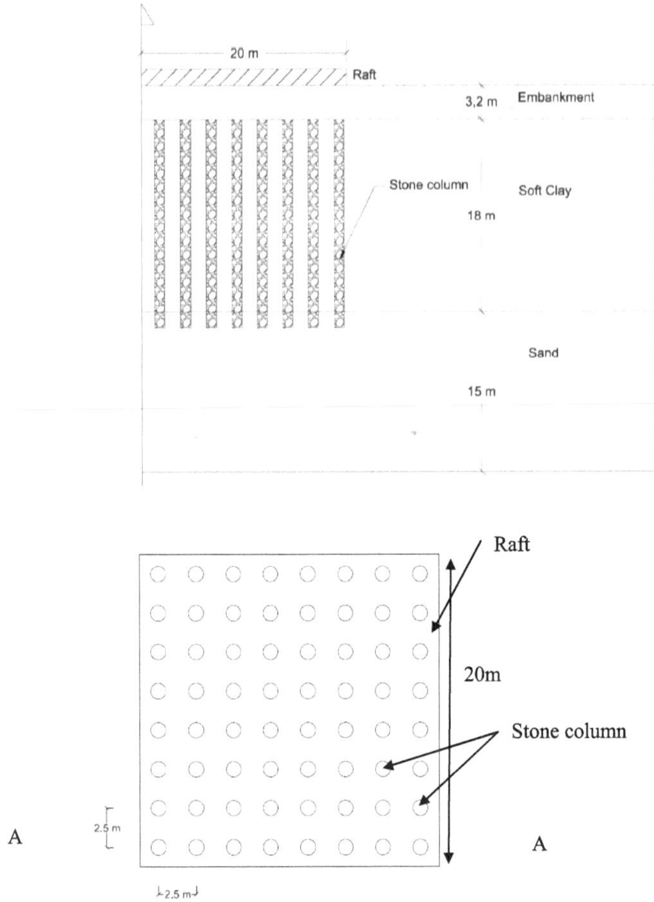

Fig. 2. Schematic of stone columns arrangement under quarter of the storage tank

The soil profile is consisting of 3 m thickness replacement layer followed by 18.2 m soft clay which rested on 15 m sand layer. The properties of the soil layers shown in Table 1. The boundary effect was examined to avoid it 60 m from the footing edge was sufficient. The model conducted on quarter of the tank. The encasement properties was presented in Table 2.

Table 1. Material Properties

	Replacement	Soft clay	Lower sand	Stone column
Top elevation	6.2	3.2	−15	3.2
Material model	Mohr-column	Soft soil	Hardening soil	Hardening soil
Drainage type	Drained	Undrained	Drained	Drained
Sat. unit weight (KN/m^3)	20	18.5	18	19
Sat. unit weight (KN/m^3)	20	18.5	18	19
$E_{50}r^{ef}$ (KN/m^2)	–	–	20,000	100E3
$E_{oed}r^{ef}$ (KN/m^2)	–	–	17,000	100E3
E_{ur}^{ef} (KN/m^2)	–	–	60,000	300E3
m	–	–	0.5	0.5
E (KN/m^2)	75,000	–	–	–
V'	0.3	–		–
v_{ur}	–	–	0.2	–
Cc	–	0.33	–	–
Cr	–	0.06	–	–
C′ (kPa)	1	3	0	1
ϕ°	42	22	36	42
OCR	–	6	1	1
Pref	–	–	100	100

Table 2. Encasement properties

Material	Geogrid
Material type	Elastic
Geosynthetic stiffness, J (KN/m)	2000

3 Results and Discussion

The results obtained from the FE model are settlements and vertical stresses on the top of the columns and the surrounding soil, variation of excess pore pressure.

3.1 Vertical Settlement

The settlement increase versus load increased. The settlement reduced by 36% when using endbearing stone column and by 33% at floating column with column length/clay thickness (L/H) equal 80% and by 28% at floating column with (L/H) equal 60%, Fig. 3. Figure 4 illustrate the improvement factors at each stage of loading. The improvement factor increased with vertical load increase. The improvement factor (Settlement treated/Settlement untreated) reach to 1.4 at the floating columns L/H = 0.6 and increased to 1.49 at L/H = 0.8. The improvement factor increased to 1.57, 1.60 at the end bearing columns and at the end bearing encased stone column. The maximum settlement value was observed at the center of the tank and the settlement reduced when

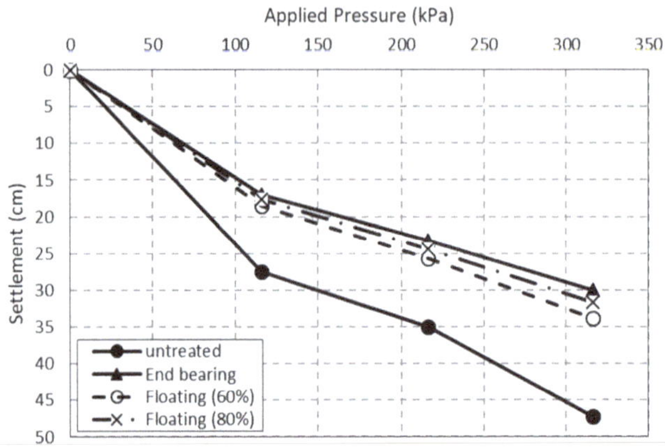

Fig. 3. Relation between vertical pressure and settlement

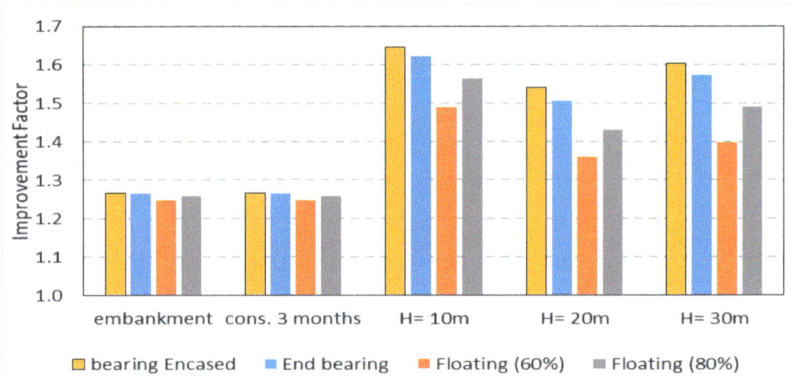

Fig. 4. Improvement factor at different stages of constructions

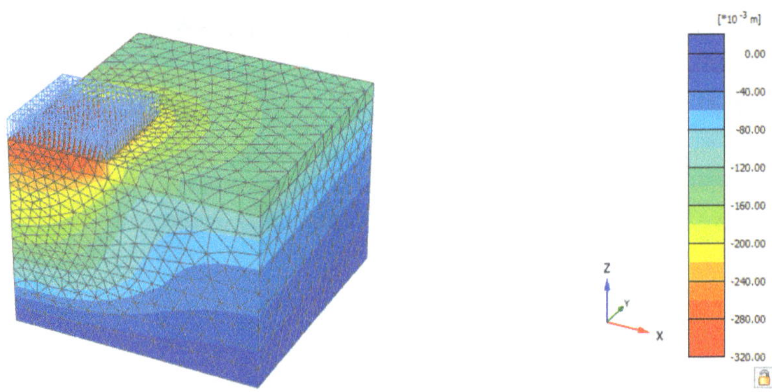

Fig. 5. Distribution of the Settlement

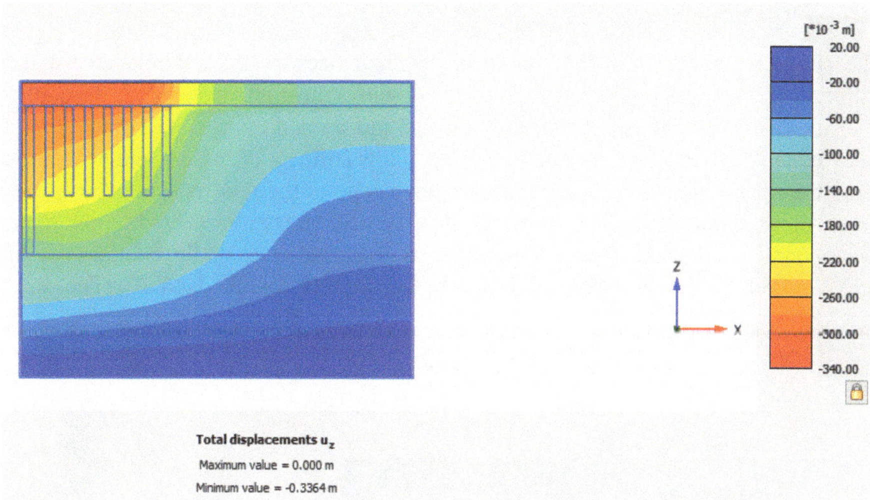

Fig. 6. Distribution of the Settlement in floating stone column (L/D = 60%) (at cross section A-A)

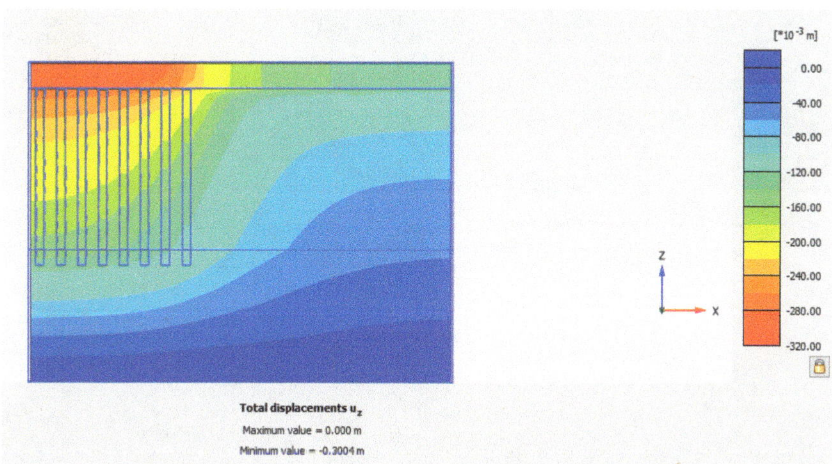

Fig. 7. Distribution of the Settlement in End bearing stone column (at cross section A-A)

the distance from the tank center increased, Fig. 5. Figures 6 and 7 shows the distribution of the settlement at cross section A-A in end bearing column case and floating (L/d = 60%) case. It is clear that the incensement of penetration length of the column improved the distribution of the settlement and reduced the settlement values.

3.2 Vertical Stresses

The stresses transfer from the soil to the column because the column have higher stiffness. The stresses on the column increased with significant values when the applied load increase. In contrast, the stresses on the soil remain approximately constant at value equal 200 kPa. This may be occur due to low stiffness of the soil and the column carry the applied load. The stress concentration ratio is the ratio between the vertical stresses on the column and the soil. The total vertical stresses measured on the top of the columns and the surrounding soil. The total vertical stresses on the columns and the stress concentration ratio decreased when the column length decrease. The stress

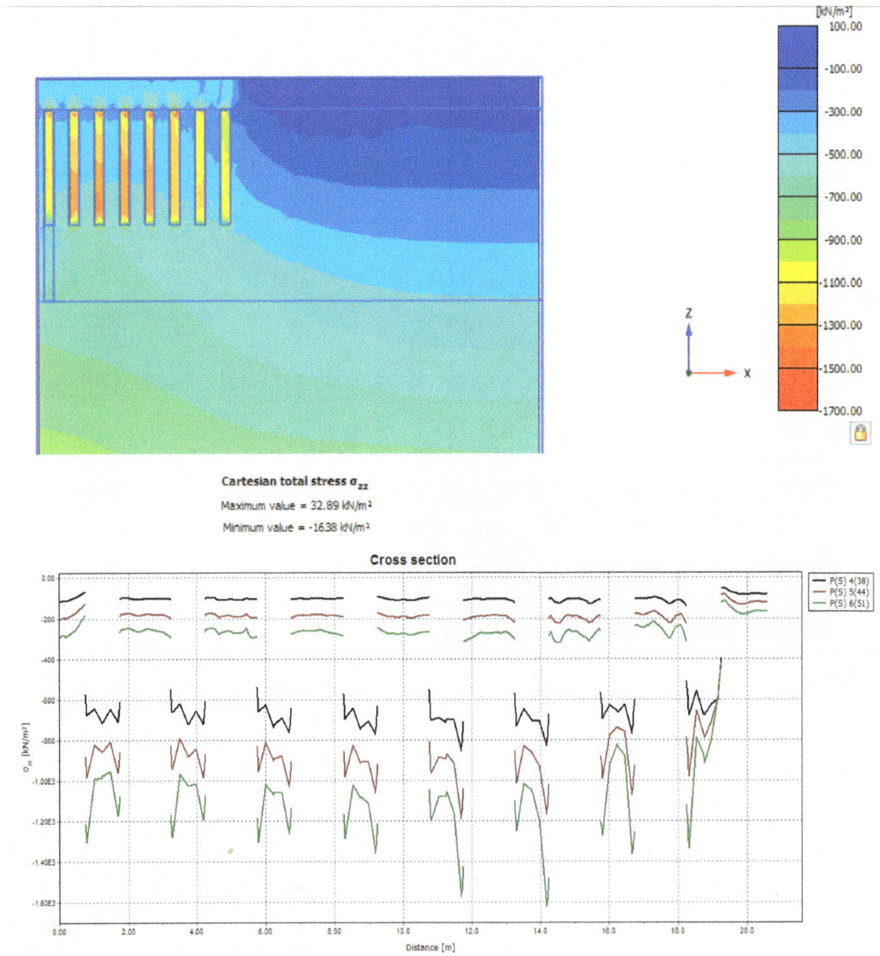

P(s) 4(38): 10m water Loadimg= 116kPa
P(s) 5(44): 20m water Loadimg= 216kPa
P(s) 6(51): 30m water Loadimg= 316kPa

Fig. 8. Distribution of the vertical stresses at floating columns L/H = 0.6 (at cross section A-A)

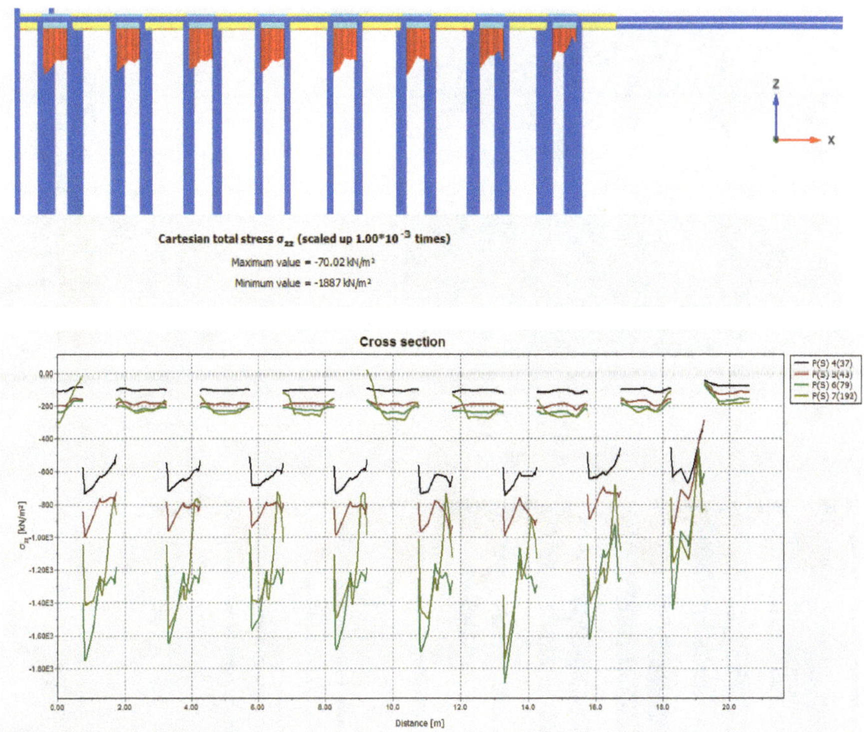

P(s) 4(37): 10m water Loadimg= 116kPa
P(s) 5(43): 20m water Loadimg= 216kPa
P(s) 6(79): 30m water Loadimg= 316kPa

Fig. 9. Distribution of the vertical stresses at floating columns L/H = 0.8 (at cross section A-A)

consolidation ratio at end of loading at L/H = 60% equal 3, Fig. 8. The stress concentration ration at the end of the loading at L/H = 80% equal 7, Fig. 9. The stress concentration ration at the end of the loading at end bearing columns equal 11, Fig. 10. The stiffness of the column increased when the column encased with geosynthetic material. The vertical stresses on the encased stone column increased and the vertical stresses on the surrounding soil decreases. The stress concentration ratio increased to 15 at the end of the loading, Fig. 11.

3.3 Excess Pore Water Pressure

The variation of excess pore water pressure is shown in Figs. 12, 13, 14, and 15. The excess pore pressure increased with loading increased. The maximum excess pore water pressure increased with decreased the column length. The maximum excess pore water pressure reduces by encasement.

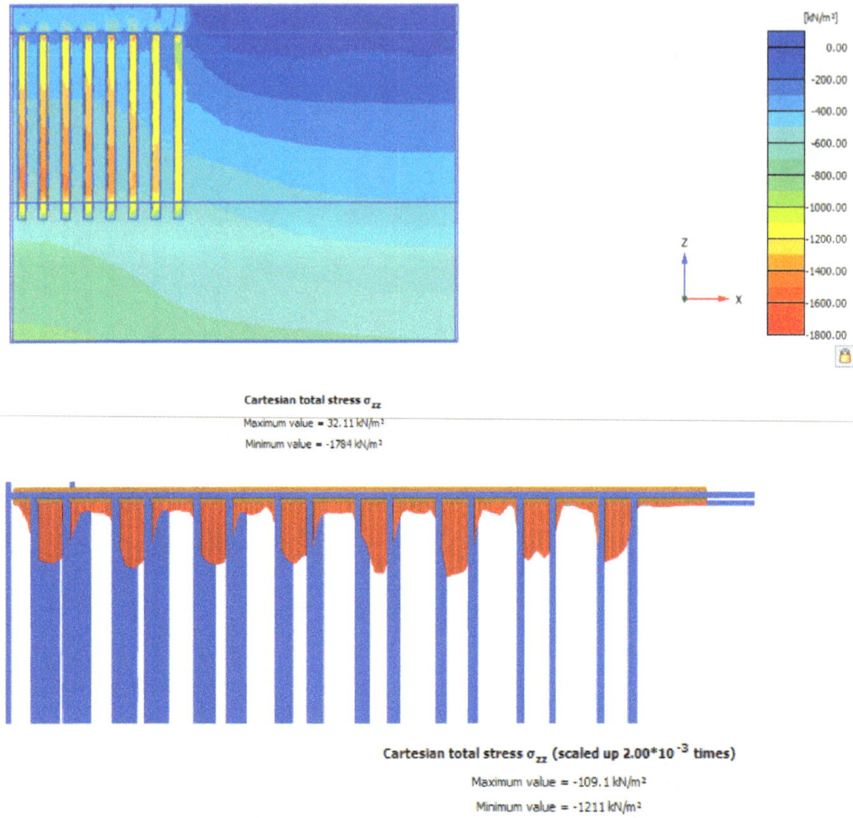

Fig. 10. Distribution of the vertical stresses at end bearing columns L/H = 1 (at cross section A-A)

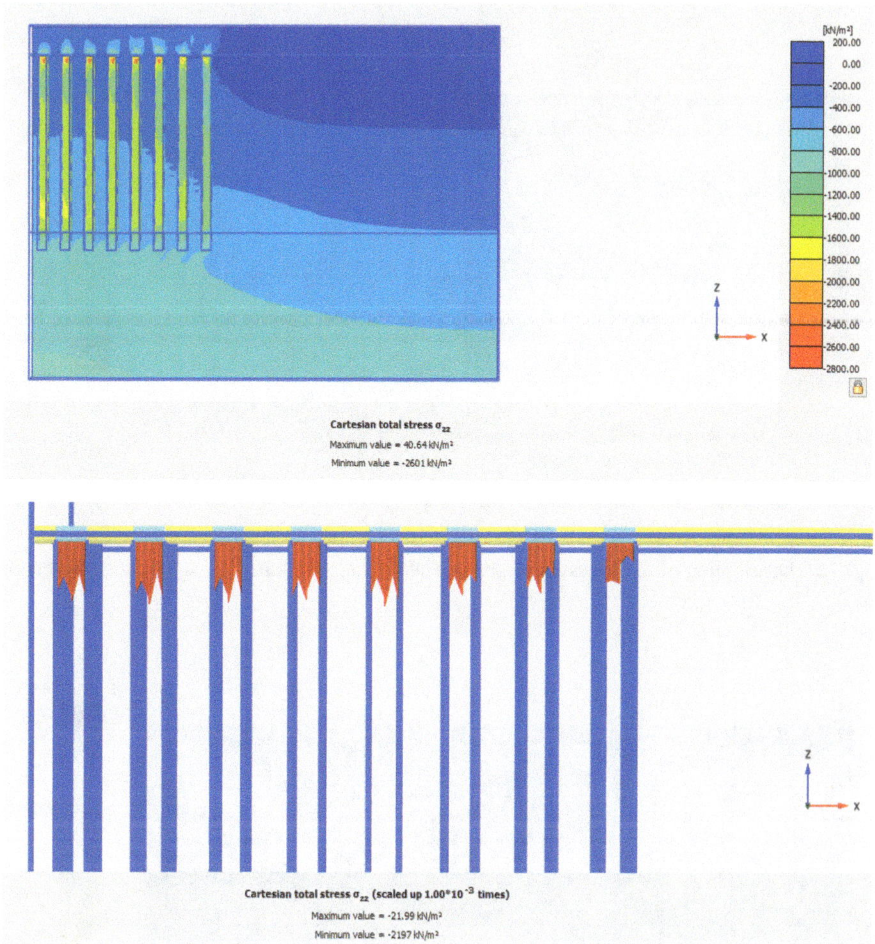

Fig. 11. Distribution of the vertical stresses at encased end bearing columns L/H = 1 (at cross section A-A)

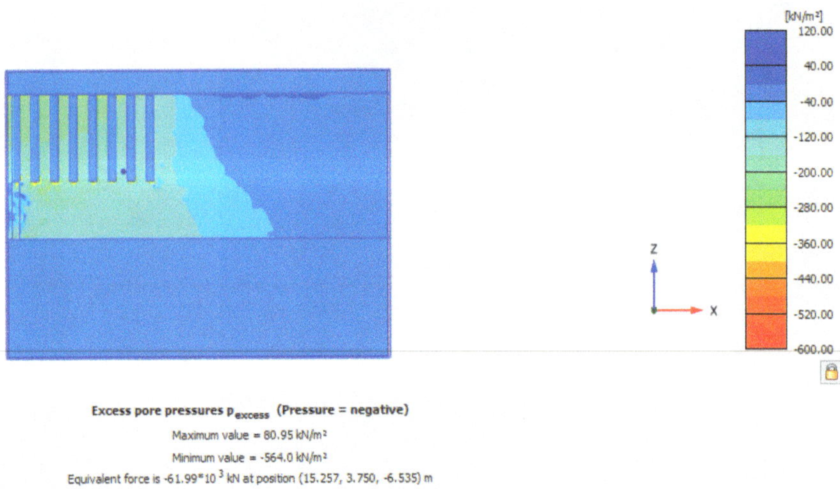

Fig. 12. Distribution of the excess pore pressure at floating columns L/H = 0.6 (at cross section A-A)

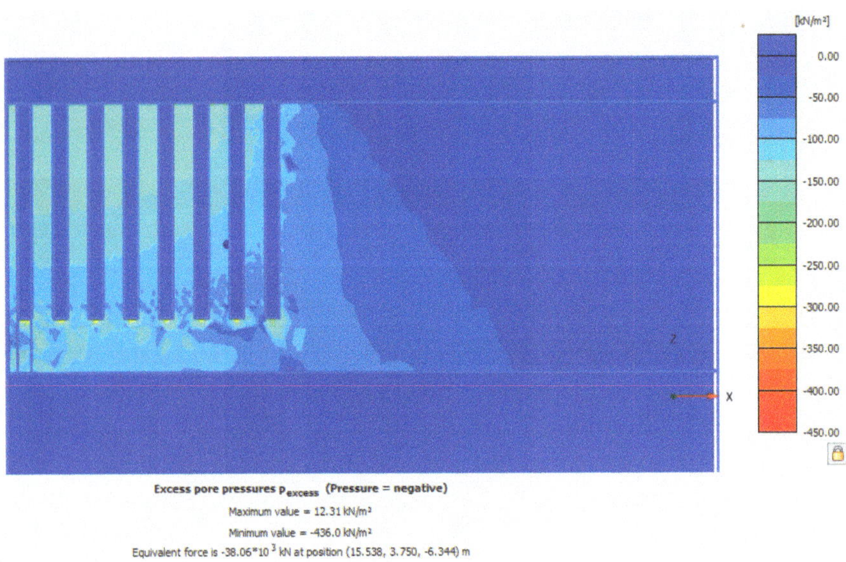

Fig. 13. Distribution of the excess pore pressure at floating columns L/H = 0.8 (at cross section A-A)

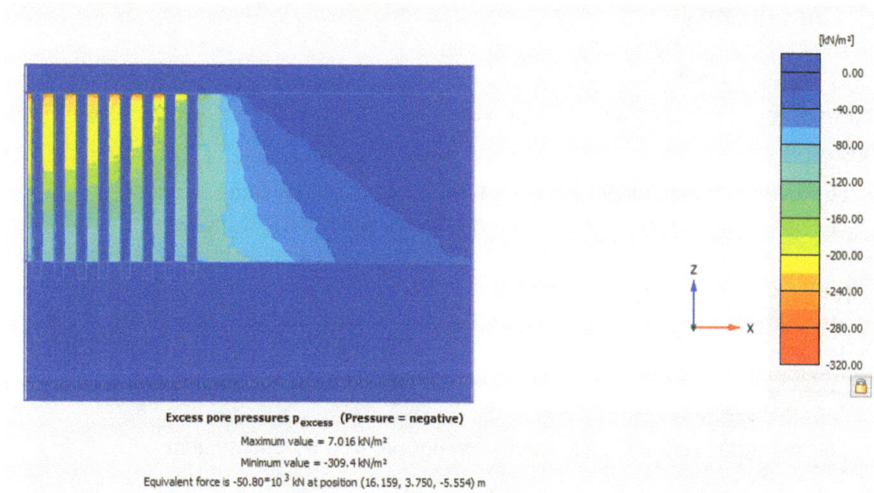

Fig. 14. Distribution of the excess pore pressure at end bearing columns L/H = 1 (at cross section A-A)

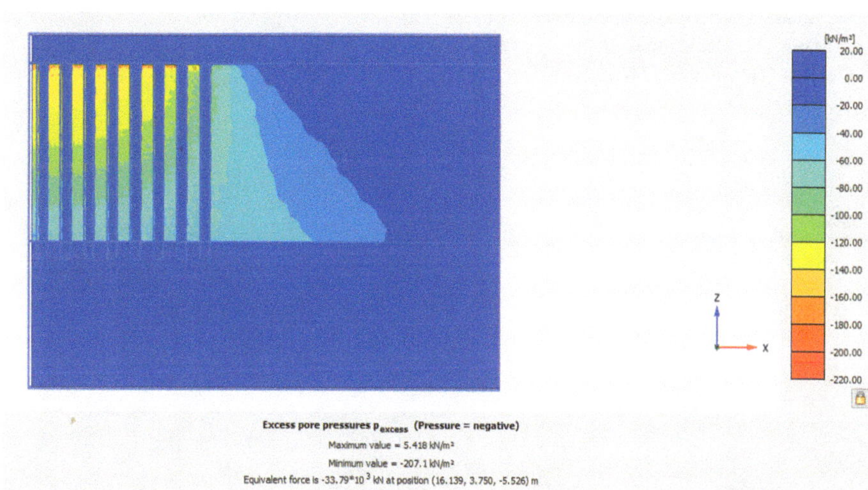

Fig. 15. Distribution of the excess pore pressure at encased end bearing columns L/H = 1 (at cross section A-A)

4 Conclusions

Finite element model was simulate water storage tank rested on the soft clay bed improved by stone columns and encased stone column. The main conclusion of this study are:

- The maximum settlement decreased with increased the column length.
- The stress concentration ratio increase with column length increased.
- The encasement increased column stiffness and decrease the excess pore water pressure and reduce the settlement.
- The vertical stresses on the column decreased when the penetration column length decreased.
- Encasement caused increased the vertical stresses on the column equal as 2.1 times the total applied vertical stresses
- The maximum excess pore water pressure reduces by encasement.

References

Gniel, J., Bouazza, A.: Improvement of soft soils using geogrid encased stone columns. Geotext. Geomembr. **27**(3), 167–175 (2009)

Murugesan, S., Rajagopal, K.: Geosynthetic-encased stone column: numerical evaluation. Geotext. Geomembr. **24**(6), 349–358 (2006)

Reliability Analysis of Stone Column Improved Soft Soil by Finite Element Approach

Anasua GuhaRay(✉) and Abhijeet Roy

Department of Civil Engineering, BITS-Pilani Hyderabad, Secunderabad 500078, India
guharay@hyderabad.bits-pilani.ac.in

Abstract. The present study considers the effect of variation of different geotechnical random variables in the design of stone column-foundation systems for assessing the bearing capacity of a highly compressible soil. The soil and stone column properties, spacing, diameter and arrangement of stone columns are considered as the random variables. A unit cell of a stone-column foundation system is modelled for assessing the bearing capacity of the system. Instead of using the conventional limit equilibrium method to calculate the probability of failure (P_f), numerical analysis is carried out in commercially available finite element (FE) software in order to simulate the mechanical behavior of the stone column. The limit state function is developed using response surface methods based on the FE models. A full factorial design is used to develop the non-linear response surface models. The study shows that the variation in cohesion of soil (c_s) is one of the most important factors influencing *FS*. It is observed that the bearing capacity of stone column improved soft ground exceeds its safe value if the coefficient of variation (*COV*) of c_s exceeds 30%. Accordingly, design guidelines for bearing capacity of improved ground, are proposed for different spacing and diameter of stone columns and geotechnical random variables.

1 Introduction

Settlement prediction and calculation is essential for assessing the foundations lying on weak soils for safety. Stone columns are very extensively and efficiently used to improve the bearing capacity of poor ground and reduce the settlement of structures built upon them. These are made of stone aggregates of sizes 20–75 mm, closely packed together in vertical trench of diameters 0.6–1.0 m and 15–20 m deep. Stone columns speed up the consolidation settlement and pore pressure dissipation by reducing the flow path length. This increases the shear strength and decreases the compressibility of compounded soil including the stone columns and neighboring soil. Other advantages of stone columns include fastening the primary consolidation settlement and restraining liquefaction (Adalier and El Gamal 2004). The load transfer theory, ultimate bearing capacity calculations and estimation of settlement of the stone columns was first proposed by Greenwood (1970).

Vesic (1972), Hughes and Withers (1974), Priebe (1976), Aboshi et al. (1979), Datye and Nagaraju (1981), Greenwood and Kirsch (1983), Van Impe and De Beer

H. Shehata and H. Poulos (Eds.): GeoMEast 2018, SUCI, pp. 39–50, 2019.
https://doi.org/10.1007/978-3-030-01917-4_3

(1983), Balaam and Poulos (1983) and Madhav et al. (1994) also later proposed similar theories. Shankar and Shroff (1997) showed from their experiments that the triangular installation pattern gave the best results. It was also proposed that the critical length for the bulging effect is nearly 3–5 times the diameter of stone columns (Mitra and Chatopadhyay 1999). Malarvizhi and Ilamparuthi (2004) concluded that the stone column encasement with geogrids results in increasing the load bearing capacity for both end bearing or floating columns. Ambily and Gandhi (2007) proposed that the maximum radial reaction of the soil against the bulging predominantly influenced the ultimate capacity of stone column and the maximum vertical movement in the stone column was restricted to about four times the diameter. McKelvey et al. (2004) designed a prototype for vibrating stone columns in a crystalline medium with properties like clay for finding the mechanisms behind the failure and made a conclusion that the long columns showed bulging failure and short columns showed punching failure. They also found that the bearing capacity of soil was not affected by columns which had length to diameter ratio greater than or equal to 6. Frikha et al. (2008) worked on the estimation of the ultimate bearing capacity of stone columns using a model which had both elastic and plastic properties. Ghazavi and Afshar (2013) examined the bearing capacity of stone columns encased with geosynthetics. Wang (2009) proposed a mathematical solution for the consolidation of stone columns reinforced soft clay foundations under loadings being time dependent. The differential equations are acquired including the smear effect and well resistance under random applied loadings like step loading, ramp loading, cyclic loading etc. Deb and Shiyamalaa (2015) developed a mathematical model for the rate of consolidation including the clogging effect. From the past literature, it can be observed that most of the reported studies are based on deterministic approach.

Soil, being a natural material, uncertainties in geotechnical engineering is unavoidable. Factor of Safety (*FS*) is chosen based on experience from past and the outcome of failure to account for this natural soil variability. This does not consider the amount of uncertainty associated with the system. Hong and Shang (1998) and Zhou et al. (1999) demonstrated that for PVD-improved ground, radial consolidation is the predominant random variable. They also concluded that the degree of consolidation is significantly affected by the type of probability distribution of random variables. Zhou et al. (1999) also worked upon achieving a specific target reliability level of degree of consolidation in a given time period. They developed a probability-based design methodology to design the PVD-improved ground. Ambily and Gandhi (2007) analysed the behavior of stone columns by finite element analysis and compared it with experimental observations. Alonso and Jimenez (2011) observed that the highest influence on the reliability results was governed by the stone column-improved ground radial coefficient of consolidation. Choobbasti et al. (2011) proposed through numerical analysis that the bearing capacity ratio for stone column improved soil in homogeneous soil was greater than heterogeneous soil. Douglas and Schaefer (2012) suggested that the Priebe's Method has approximately 89% probability that the measured settlement will be less than the estimated settlement. Fattah and Majeed (2012) proposed that stone columns increase the bearing capacity in soft soils, reason being they are stiffer than the replaced material and these columns provide shearing resistance giving vertical support to overlaying structures or large embankments. Bari et al. (2013) experimented on the reliability analysis of soil consolidation with PVD by taking into consideration the inherent spatial variability of soil properties. Bari and Shahin (2014) suggested a simplified probabilistic method in which the inherent variability of the coefficient of consolidation of the soil is considered. Deb and Majee (2014) estimated the diameter and spacing for stone column improved ground using the probability based design charts.

From the above literature, it is observed that most of the reliability analysis of soil consolidation via vertical drain is conducted on PVD-improved ground. Limited studies are conducted on stone column-improved ground. The present paper studies the effect of variation of four different geotechnical random variables on bearing capacity and consolidation settlement of stone columns. Probability of failure (P_f) is analysed for three target safe loads of 200 kN, 300 kN and 400 kN. Instead of using the conventional LEM to calculate P_f, numerical analysis is carried out in commercially available FE software in order to simulate the mechanical behavior of the stone column. The limit state function is developed using response surface methods based on the FE models. A full factorial design is used to develop the response surface models. P_f is obtained from the limit state function by applying the basic definition of reliability. The analysis is carried out in commercially available software Matlab R2016a. Finally, design guidelines are proposed for different dimension of stone columns and variations of geotechnical random variables, which may lead to a safe and economic design.

2 Finite Element Modeling of Stone Columns

The limit equilibrium methods for analysis of stone columns are still more popular due to its simplicity. But the elasto-plastic behavior of soil is not considered in this method. The principal difference between these two approaches is that the *LEM* is based on the static of equilibrium, whereas *FEM* utilizes the stress-strain relationship or constitutive law in its formulation. In the present study, the stone column is modeled in commercially available software PLAXIS 2D 2017. An initial condition with geostatic stress is defined in the present model. Geostatic stress condition is considered to simulate the earth pressure condition. Lateral earth pressure coefficient at rest (K_0) is defined using Jacky's formula

$$K_o = 1 - \sin\phi$$

where φ is the angle of internal friction of the corresponding soil layer.

Horizontal fixity is provided to the vertical edge of the axisymmetric model. In the bottom edge of the model, both vertical and horizontal fixity are applied as the bottom is assumed to be non-yielding (Fig. 1). The vertical edge is kept at a sufficient distance so that the column behavior is not affected due to the side edges. In the numerical study, a strain controlled loading is applied in increments. The strain controlled loading is applied in such a manner so that with each load step, the soil is pushed by a displacement of 0.001 m per second. A plate is placed over the surface of the unit cell.

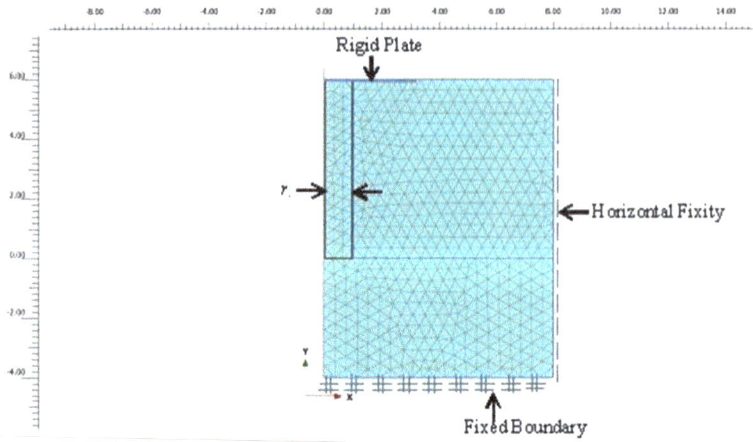

Fig. 1. Finite element model of unit cell of stone column

The plate acts as a loading platform and prevents differential settlements between the surfaces of the column and the surrounding soil.

The diameter ratio (N) of stone columns is considered in the range of 2–6. A constant depth of 6 m is considered with spacing (S) varying from 1 to 4 m. Table 1 summarises the different cases considered depending on the values of N and S for triangular pattern of arrangement. d_c and D_e are the diameter and equivalent diameter of stone column respectively. For a triangular pattern of arrangement of stone columns, $D_e = 1.05S$. According to Mitchell (1981), the safe superstructure load ($Q_{superstructure}$) for stone columns in soft to medium stiff clays lies in the range of 200–300 kN. In the present study, three different target safe superstructure loads are considered (200 kN, 300 kN and 400 kN).

Table 1. Triangular distribution of stone columns

Case	S/d_c	S (m)	d_c (m)	$D_e = 1.05S$	$N = D_e/d_c$
I	2.38	1.0	0.42	1.05	2.5
II	3.33	1.5	0.45	1.58	3.5
III	4.29	2.0	0.47	2.10	4.5
IV	5.55	2.5	0.45	2.63	5.5
V	4.54	3.0	0.66	3.15	4.5
VI	3.33	3.5	1.05	3.68	3.5
VII	2.38	4.0	1.68	4.20	2.5

2.1 Soil

For geo materials like soils and rocks, in principal both the cohesion and the friction angle may increase (hardening) or decrease (softening) with the progress of plastic deformation and that the hardening/softening behaviors are related only to the increase/decrease of the cohesion while the internal friction remains constant during the plastic deformation (Chen et al. 2012). The simple linear elastic–perfectly plastic Mohr–Coulomb model is used to model the behavior of soil in order to be consistent with the conventional foundation design (Potts and Zdravkovic 2001). The model is used as a first approximation of soil behavior in general. A constant average stiffness is estimated for the soil layer. Due to this constant stiffness, computations tend to be relatively fast and a first estimate of deformations can be obtained. In general stress state, the model's stress-strain behaves linearly in the elastic range, with two defining parameters from Hooke's law (Young's modulus, E and Poisson's ratio, μ). There are two parameters which define the failure criteria (the friction angle, φ and cohesion, c) and also a parameter to describe the flow rule (dilatancy angle, ψ) which comes from the use of non-associated flow rule and is used to model a realistic irreversible change in volume due to shearing.

To model the soil, 15-noded triangular elements with the 12-point integration rule are used. The properties of soil and stone column material are provided in Table 2. The material properties are given for total stress analysis. In Table 2, *s* denotes soil and *c* denotes stone column material. The mean values of the geotechnical random variables, their coefficient of variation (*COV*) and statistical distribution are considered from past research works, in absence of any site data. The modulus of elasticity of the clay is assumed to be $E_s = c_s \times 250$ (E = 200–500 × c_s) (Bowles 1996). *FS* obtained by deterministic analysis corresponding to different target safe loads are summarized in Table 3. It can be seen from Table 3 that *FS* against bearing capacity increases with increase in spacing and diameter ratio. Minimum *FS* is observed for a spacing of 1 m and diameter ratio of 2.5. A unit cell approach is considered for the present study.

Table 2. Statistics of input parameters

Variables	Mean μ	*COV* (%)	Distribution
φ_s (°)	15	5–20	Log-normal
c_s (kN/m^2)	25	20–50	Log-normal
γ_s (kN/m^3)	20	7	Gaussian
φ_c (°)	38	5–20	Log-normal
E_s(kN/m^2)	300c_s	20–50	Log-normal
E_c(kN/m^2)	30000	30	Log-normal
μ_s	0.4	–	–
μ_c	0.2	–	–

2.2 Model Discretization

The process of dividing a finite element numerical model into smaller pieces is known as discretization or meshing. Very fine global coarseness with global refinement (to increase the number of elements globally) and cluster refinement (to increase the

Table 3. Deterministic factors of safety for bearing capacity

Cases	*FS* for		
	Q = 200 kN	Q = 300 kN	Q = 400 kN
I	2.22	1.48	1.11
II	4.09	2.72	2.04
III	6.59	4.39	3.29
IV	9.78	6.52	4.89
V	14.98	9.98	7.49
VI	22.99	15.33	11.49
VII	39.46	26.31	19.73

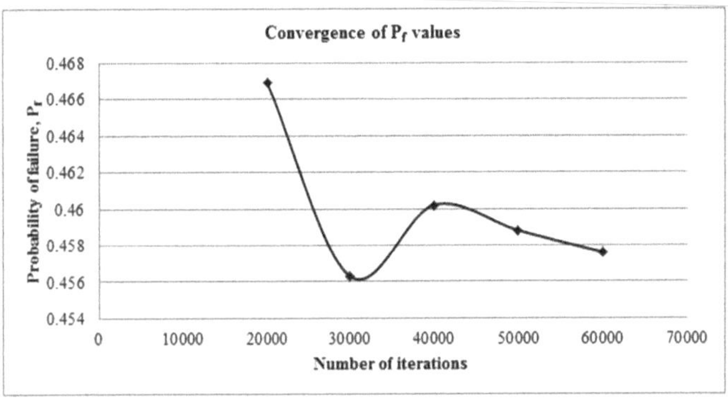

Fig. 2. Mesh convergence study of FE model

number of elements in a particular cluster) is used for meshing. Line refinement (to increase the element numbers at the cluster boundaries) is done on the footing. For the present study, since the settlement of the clay around the stone columns is of prime importance, a denser mesh was generated in the soil zone surrounding the stone column. There are 1917 elements and 15,619 nodes in the model. A mesh convergence study is also performed to arrive at the optimum size of meshing (Fig. 2).

PLAXIS defines R_{inter} as the strength reduction factor for interfaces between the soil and the stone column. R_{inter} is taken as 0.8. At the corners between the soil and the column, the interface value R_{inter} is chosen as 1, because when interfaces are extended below or around corners of structures to avoid stress oscillations, these extended interfaces are not meant to model soil–structure interaction behavior, but just to allow sufficient flexibility. Hence, $R_{inter} < 1$ for these conditions may introduce an unrealistic strength reduction in the ground, which may lead to unrealistic soil behaviour or even failure.

3 Design of Experiments for *FS* of the Stone Column

A factorial experiment is an experimental procedure in which design variables are varied together, instead of one at a time, to investigate the joint effects of the factors on a response variable. Factorial design is an important special case in which each of the k factors of interest has only two levels. In this experimental design 2^k = 24 = 16 sets of data have been generated for six input parameters, the points being estimated as $\mu + 1.96\sigma$ and $\mu - 1.96\sigma$. Analysis is shown for a target safe load of 200 kN (Table 4).

Regression analysis is carried out by these *FS* values obtained by FEM to get a multiple non-linear response surface model (Table 5). Correlation coefficients are set to zero for conservative deformation behavior, and a multiple non-linear correlation model between the geotechnical parameters is assumed.

Table 4. FS Corresponding to 16 sample points by FEM

Statistics	γ_s	φ_s	c	φ_c	Diameters of columns (d_c)					
					0.42	0.45	0.47	0.66	1.05	1.68
Mean μ	20	15	25	38	1.11	2.04	3.29	7.49	11.49	19.73
COV	7	10	10	10	–	–	–	–	–	–
SD σ	1.40	1.50	2.5	3.8	–	–	–	–	–	–
$\mu + 1.96\sigma$	22.74	17.94	29.9	45.45	–	–	–	–	–	–
$\mu - 1.96\sigma$	17.26	12.06	20.10	30.55	–	–	–	–	–	–
1	22.74	17.94	29.9	45.45	1.18	1.09	1.21	1.16	1.19	1.28
2	22.74	17.94	29.9	30.55	0.88	0.90	0.85	0.90	0.96	1.29
3	22.74	17.94	20.10	45.45	1.02	0.97	0.96	1.05	1.04	1.10
4	22.74	17.94	20.10	30.55	0.98	0.85	0.84	0.87	0.86	0.90
5	22.74	12.06	29.9	45.45	0.96	0.93	0.93	0.95	0.96	1.02
6	22.74	12.06	29.9	30.55	0.98	0.82	0.94	0.96	0.87	0.94
7	22.74	12.06	20.10	45.45	1.20	1.13	1.14	1.158	1.171	1.265
8	22.74	12.06	20.10	30.55	0.97	0.85	0.86	0.94	0.94	1.07
9	17.26	17.94	29.9	45.45	0.98	0.95	0.94	0.99	0.96	1.01
10	17.26	17.94	29.9	30.55	1.09	1.05	1.08	1.11	1.11	1.11
11	17.26	17.94	20.10	45.45	0.95	0.86	0.83	0.96	0.87	0.94
12	17.26	17.94	20.10	30.55	0.97	0.89	0.94	0.94	0.96	0.97
13	17.26	12.06	29.9	45.45	0.97	0.84	0.85	0.99	0.98	0.99
14	17.26	12.06	29.9	30.55	1.02	0.99	0.98	1.04	0.99	1.02
15	17.26	12.06	20.10	45.45	0.97	0.94	0.97	0.98	0.97	0.98
16	17.26	12.06	20.10	30.55	1.07	1.03	1.11	1.11	1.05	1.09

Table 5. Response surface model for different diameters

d_c (m)	Multiple non-linear equation
0.42	$-1.007 - 0.51\mathbf{e}^{-0.002\gamma} + 0.3397\ln(\mathbf{c_s}) + 0.00859(\phi_\mathbf{c}) + 0.7068\mathbf{e}^{0.0072(\phi_s)}$
0.45	$-1.148 - 0.06679e^{-0.002\gamma} + 0.57\ln(c_s) + 0.00837(\phi_c) + 0.93e^{0.0072(\phi_s)}$
0.47	$-2.1785 - 0.669e^{-0.002\gamma} + 0.61\ln(c_s) + 0.0114(\phi_c) + 1.12e^{0.0072(\phi_s)}$
0.66	$-1.6487 - 0.8839e^{-0.002\gamma} + 0.4155\ln(c_s) + 0.0098(\phi_c) + 0.8215e^{0.0072(\phi_s)}$
1.05	$-1.1098 - 0.0151e^{-0.002\gamma} + 0.58\ln(c_s) + 0.00588(\phi_c) + 0.937e^{0.0072(\phi_s)}$
1.68	$0.2367 - 1.3224e^{-0.002\gamma} + 0.8656\ln(c_s) + 0.00258(\phi_c) + 0.647e^{0.0072(\phi_s)}$

The general expression for factor of safety as a function of diameter of stone columns is given as follows

$$FS = K + Ae^{-0.002\gamma} + B\ln(c_s) + D(\phi_c) + Ee^{0.0072(\phi_s)}$$

Where,

$K = -5.40d_c^3 + 18.32d_c^2 - 17.36d_c + 3.18$
$A = 5.88d_c^3 - 19.84d_c^2 + 18.63d_c - 4.50$
$B = 0.370e^{0.48} * d_c$
$C = -0.0057 * d_c + 0.012$
$D = -0.38 * d_c^2 + 0.61d_c + 0.70$

The performance function is defined as $f(x) = FS - 1$

The FS values are fitted to a proper statistical distribution using MATLAB Statistical Toolbox (MATLAB R2016a) and the P_f value is obtained by First Order Second Moment Method

Case 1: Effect of variation of φ_c

P_f for bearing capacity failure obtained for different spacing and diameter ratio of stone columns, corresponding to different variations of φ_c, are shown in Fig. 3a. From the figure, it is quite evident that when $S = 1$ m and $N = 2.5$, the P_f is above the acceptable limit for any variation of φ_c. Figure 3b shows the variations of P_f for the three different target loads for $S = 1$ m and $N = 2.5$. It may be concluded that for the assumed mean values of the geotechnical random variables, the stone column can sustain a maximum safe load of 200 kN, considering any variation of the random variables and any arrangement of the stone columns.

Case 2: Effect of variation of φ_s

Secondly, the effect of variation of internal angle of friction of soil (φ_s) on bearing capacity failure is analyzed (Fig. 4). It is noted from the figure, that P_f shows a similar variation as in Case 1. This shows that φ_s and φ_c have almost similar effect on the bearing capacity of a stone column improved ground.

Case 3: Effect of variation of c_s

The cohesion of soil (c_s) affects the bearing capacity of a stone column improved ground significantly. Figure 5 demonstrates the P_f for different spacing and diameter

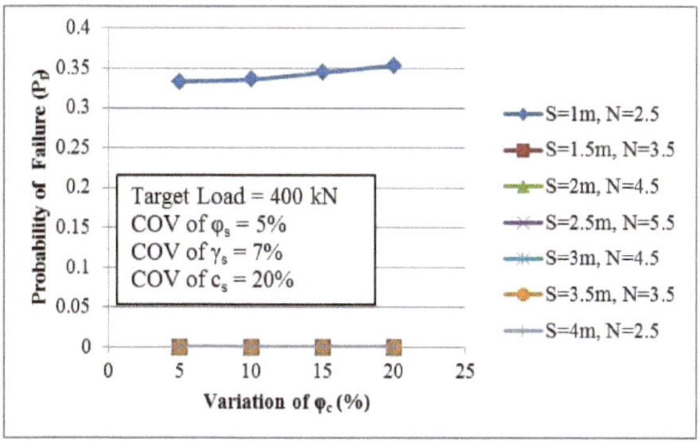

(a) P_f for different variations of φ_c for different spacing and diameter ratios

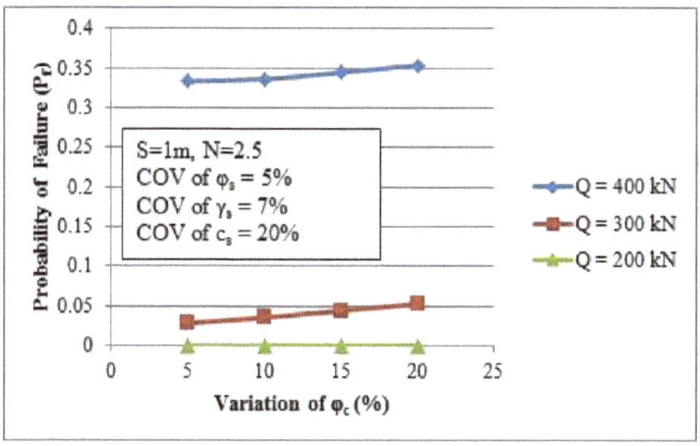

(b) P_f for different variations of φ_c for different target loads

Fig. 3. **(a)** P_f for different variations of φ_c for different spacing and diameter ratios. **(b)** P_f for different variations of φ_c for different target loads

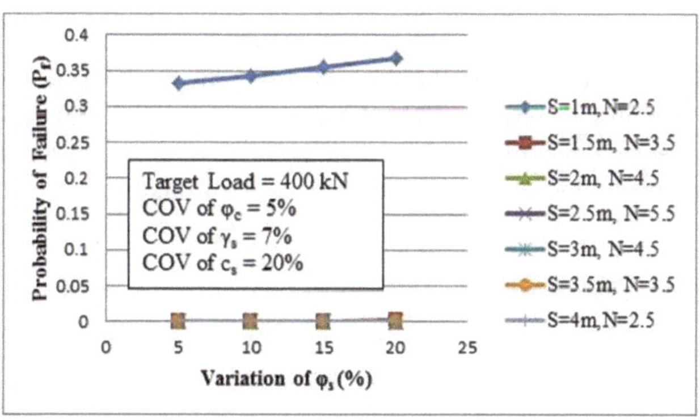

Fig. 4. P_f for different variations of φ_s for different spacing and diameter ratios

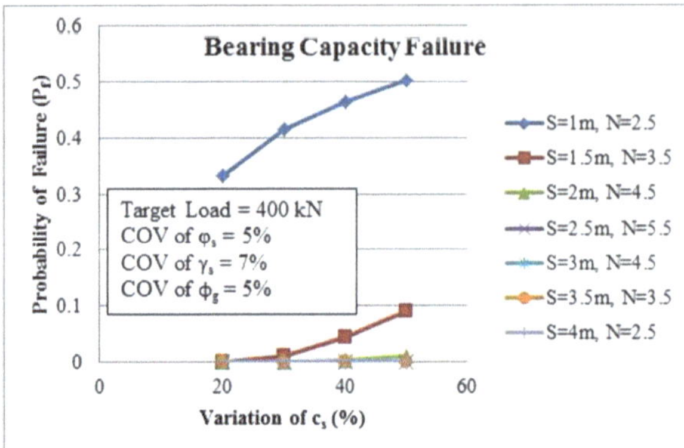

Fig. 5. P_f for different variations of c_s for different spacing and diameter ratios

ratios corresponding to different variations on c_s. It can be inferred from the figure that when *COV* of c_s exceeds 20%, the P_f crosses the acceptable limits when the spacing and diameter ratio of stone columns are less than 1.5 m and 3.5 respectively.

4 Conclusion

It is well known that the variation in geotechnical properties from the estimated value is the rule rather than the exception. The factor of safety concept was developed initially to account for those variations in design. In the present study, the effect of variation of geotechnical random variables on bearing capacity of stone column improved ground is studied by a finite element approach. The study shows that the variation in cohesion of soil (c_s) is one of the most important factors influencing *FS*. It is observed that the bearing capacity of stone column improved soft ground exceeds its safe value if the coefficient of variation (*COV*) of c_s exceeds 30%. Accordingly, design guidelines for bearing capacity of improved ground, are proposed for different spacing and diameter of stone columns and geotechnical random variables. The major hindrance behind the popularity of the reliability based design approach lies in the lack of knowledge inherent in statistical methods. The design guidelines provided for penetration depth for both types of backfill may prove useful to practitioners. Additionally, for sensitive and big-budget projects, reliability analysis can help in making appropriate decisions without compromising the economy and safety of the structures.

References

Aboshi, H., et al.: The compozer - a method to improve characteristics of soft clays by inclusion of large diameter sand columns. In: Proceedings of International Conference on Soil Reinforcement: Reinforced Earth and Other Techniques, vol. 1, pp. 211–216 (1979)

Adalier, K., Elgamal, A.: Mitigation of liquefaction and associated ground deformations by stone columns. Eng. Geol. **72**(3), 275–291 (2004)

Alonso, J., Jimenez, R.: Reliability analysis of stone columns for ground improvement. In: Geo-risk, ASCE, pp. 493–500 (2011)
Ambily, A.P., Gandhi, S.R.: Behaviour of stone columns based on experimental and FEM analysis. J. Geotech. Geoenviron. Eng. **133**, 405–415 (2007)
Balaam, N.P., Poulos, H.G.: The behavior of foundations supported by clay stabilized by stone columns. In: Proceedings of Specialty sessions, VII European Conference on Soil Mechanics and Foundation Engineering, Helinski, vol. 2 (1983)
Bari, W., Shahin, A., Nikraz, H.R.: Probabilistic analysis of soil consolidation via prefabricated vertical drains. Int. J. Geomech. ASCE **13**(6), 877–881 (2013)
Bari, M.W., Shahin, M.A.: Probabilistic design of ground improvement by vertical drains for soil of spatially variable coefficient of consolidation. Geotext. Geomembr. **42**(1), 1–14 (2014)
Bowles, J.E.: Foundation Analysis and Design, 5th edn., p. 1175. McGraw-Hill, New York (1996)
Chen, S.L., Abousleiman, Y.N., Muraleetharan, K.K.: Closed-form elastoplastic solution for the wellbore problem in strain hardening/softening rock formations. Int. J. Geomech. ASCE **12**, 494–507 (2012)
Choobbasti, A.J., Zahmatkesh, A., Noorzad, R.: Performance of stone columns in soft clay: numerical evaluation. Geotech. Geol. Eng. **29**(5), 675–684 (2011)
Datye, K.R, Nagaraju, S.S: Design approach and field control for stone columns. In: Proceedings of 10th International conference on Soil Mechanics and Foundation Engineering, Stockholm, pp. 637–640 (1981)
Deb, K., Majee, A.: Probability-based design charts for stone column-improved ground. Geomech. Eng. **7**(5), 539–552 (2014)
Deb, K., Shiyamalaa, S.: Effect of clogging on rate of consolidation of stone column–improved ground by considering particle migration. Int. J. Geomech. ASCE (2015). https://doi.org/10.1061/(ASCE)GM.1943-5622.0000492,04015017
Douglas, S.C., Schaefer, V.R.: Reliability of the Priebe method for estimating stiffness. Gr. Improv. **167**(G12), 108–121 (2012)
Fattah, M.Y., Majeed, Q.G.: Finite element analysis of geogrid encased stone columns. Geotech. Geol. Eng. **30**, 713 (2012). https://doi.org/10.1007/s10706-011-9488-8
Frikha, W., Bouassida, M., Canou, J.: Calibration of an elastoplastic model for theprediction of stone column ultimate bearing capacity. In: GeoCongress 2008: Geosustainability and Geohazard Mitigation, pp. 604–611 (2008)
Ghazavi, M., Afshar, J.N.: Bearing capacity of geo synthetic encased stone columns. Geotext. Geomembr. **38**, 26–36 (2013)
Greenwood, D.A.: Mechanical improvement of soils below ground surface. In: Proceedings of Conference on Ground Engineering Institution of Civil Engineers, London, pp. 11–22 (1970)
Greenwood, D.A., Kirsch, K.: Specialist ground treatment by vibratory and dynamic methods – state of art. In: Advances in Piling and Ground Treatment for Foundations, Institution of Civil Engineers, London, pp. 17–45 (1983)
Hong, H.P., Shang, J.Q.: Probabilistic analysis of consolidation with prefabricated vertical drains for soil improvement. Can. Geotech. J. **35**(4), 666–677 (1998)
Hughes, J.M.O., Withers, N.J.: Reinforcing of soft cohesive soils with stone columns. Gr. Eng. **7** (3), 42–49 (1974)
Madhav, M.R., Alamghir, M., Miur, N.: Improving granular column capacity by geogrid reinforcement. In: Proceedings of 5th International Conference on Geotextile, Geomembranes and Related Products, Singapore, pp. 351–356 (1994)
Malarvizhi, N., Ilamparuthi, K.: Load versus settlement of clay bed stabilized with stone and reinforced stone column. In: 3rd Asian Regional Conference on Geosynthetic, pp. 322–329 (2004)

McKelvey, D., Sivakumar, V., Bell, A., Graham, J.: Modelling vibrated stone columns in soft clay. Proc. Inst. Civ. Eng. Geotech. Eng. **157**(GE3), 137–149 (2004)
Mitra, S., Chathpadhyay, B.C.: Stone columns and design limitations. In: Proceedings of Indian Geotechnical Conference, pp. 201–205 (1999)
Potts, D.M., Zdravkovic, L.: Finite Element Analysis in Geotechnical Engineering: Application. Thomas Telford, London (2001)
Priebe, H.J.: Evaluation of the settlement reduction of a foundation improved by Vibro replacement. Bautechnik **5**, 160–162 (1976)
Sankar, K., Shroff, A.V.: Experimental study on floated stone column in soft kaolinite clay. In: Proceedings of Indian Geotechnical Conference, held at Vadodara, pp. 265–268 (1997)
Van Impe, W.Y., De Beer, E.: Improvement of settlement behavior of softy layers by means of stone columns. In: Proceedings of 8th International Conference on Soil Mechanics and Foundation Engineering, Helsinki, pp. 309–312 (1983)
Vesic, A.S.: Expansion of cavities in an infinite soil mass. J. Soil Mech. Found. Eng. ASCE **98** (SM3), 265–290 (1972)
Wang, G.: Consolidation of soft clay foundation reinforced by stone columns under time-dependent loading. J. Geotech. Geoenviron. Eng. ASCE **135**(12), 1922–1931 (2009)
Zhou, W., Hong, H.P., Shang, J.Q.: Probabilistic design method of prefabricated vertical drains for soil improvement. J. Geotech. Geoenviron. Eng. **125**(8), 659–664 (1999)

Enhancing the Soft Clay Bearing Capacity by Using Cement Dust – Lime Columns

Adel Belal[1], El-Mashad[2], S. A. Elsoud[1], and M. M. AbdelMonem[3](✉)

[1] College of Engineering and Technology (Cairo), AASTMT, Cairo, Egypt
[2] Soil Mechanics Department, Construction Research Institute (CRI), National Water Research Center (NWRC), Cairo, Egypt
[3] Construction Research Institute (CRI), National Water Research Center (NWRC), Cairo, Egypt
mohamedmedhat17@hotmail.com

Abstract. Soft clay deposits cause a lot of problems to structures due to low shear strength and high deformation. Soft clay properties can be improved by several methods such as; prefabricated vertical drains, sand cushion reinforced with shredded waste tire, geofoam, stabilization with lime silica and silica fume and stabilization with flyash. The proposed research studies the effect of adding encased cement dust and lime mix column to soft clay on the clay properties such as bearing capacity and settlement. Model tests were performed in different diameters (3.5 cm, 4.0 cm, and 5.0 cm) and different lengths (60%, 80%, and 100%) of the clay layer depth. Non-woven geotextile was used to confining the mix. Some laboratory tests were done on the geotextile, lime, cement dust and clay that artificially prepared from kaolinit. The model was performed on soil that has undrained shear strength of 7 kPa. LVDT was installed to measure the corresponding settlement to each load. A series of 9 tests were carried out using columns considering the case of no columns. A relationship between stress and settlement were plotted and finally the improvement percentages were found. It was found that; the increasing in column diameter and length improve the bearing capacity and reduce the settlement, the floating column gives a better improvement than the end bearing column. The cement dust – lime column enhance the soft clay bearing capacity by 353%.

Keywords: Cement dust – lime column · Bearing capacity
Ground modification · Geotextile · Soil reinforced · Floating · Soft clay
Laboratory test

1 Introduction

The structures that constructed on soft soils face many problems such as large settlement and low bearing capacity. The traditional treatments are very expensive due to raw material prices and the machines cost for executing the project. To solve these problems, a method that features economy and give an acceptable improvement must be used to enable the government to continue the development process. There are many methods that can be used to improve soft clay properties such as pre-consolidation using prefabricated vertical drains, stone (or granular) columns (Hughes and Withers

H. Shehata and H. Poulos (Eds.): GeoMEast 2018, SUCI, pp. 51–65, 2019.
https://doi.org/10.1007/978-3-030-01917-4_4

1974), grouting with lime and lime silica (Fattah et al. 2015) and trench of cement dust (Yaseen et al. 2009). Columns insertions in soft formation are used widely because of its good improvement of soft soils. The most common columns used are granular column, geosynthetic encased column, concrete pile, lime pile and cement-lime columns. The cement dust - lime column is one of the column types that used to improve the shear strength of the soil and reduce the settlement of the soil. It is considered an economic method and gives an acceptable improvement than any other methods such as concrete piles as the concrete pile transfer the stress applied to the stronger layers under the soft soil, that mean the length, diameter and design consideration can control the cost of the piles. Cement dust can improve the soft clay bearing capacity by consuming the clay water content in hydration process. Many researches were done on cement dust. Khdhair (1997) used one type of stabilization which was carried out on Nasriyia sand dunes by cement dust, he made an experimental set up consisting of a wind tunnel that was designed and manufactured, and tests were performed to check the suitability of this material. Three different suspensions were prepared at a constant concentration with different mix ratios of cement dust and Portland cement. The satisfactory results were obtained from using cement in a ratio of (90% cement dust +10% Portland cement) suspended in water in concentration of 0.33 kg of this material for each liter of water applied to sand at $10\ell/2\ m$ rate. Numbers of cement dust and clay-type soils were used by McCoy and Kriner (1971) to study the soil stabilization, soil-cement dust mixes containing 3, 8, and 10% of CKD were tested for various engineering properties, such as the unconfined compressive strength, moisture-density relationship, liquid limits (LL), plastic limit (PL), plasticity index (PI), and shrinkage limit. The study found that the use of cement dust improve the soils for sub base application. Azad (2000) found that the compressive strength of the soil increased in the presence of cement dust and also the plasticity index of the soil increased.

2 Materials

2.1 Kaolinit (Clay)

In current study, Clay was prepared from kaolinit. This material is an artificial material. It is powder in dry case. When kaolinit material mixed with water, the powder became clay material so it can be used to simulate soft clay. Some tests were performed on clay and (Table 1) shows the clay properties.

2.2 Sand

Sand was used with 66% because the mix is very fine due to the presence cement dust and lime. Grain size distribution test was performed on the sand material. Figure 1 shows the grain size distribution of sand. The tests showed that the classification of the sand was fine to coarse sand, trace of gravel and trace of silt. Table 2 shows the classification of the sand used. Some other tests were performed on sand samples such as compaction and relative density. Table 3 showed the properties of the sand used.

Table 1. Tests performed on kaolinit and results.

Test type	Property	Value	Standard
Atterberg limit	Liquid limit	29.2%	Egyptian code for soil mechanics and design and executing the foundation 2009, part 2
	Plastic limit	18.1%	
	Plasisty index	11.1%	
	Unified soil classification	CL	
Consolidation test	Cc	0.0562	
	Cr	0.0079	
	eo	1.05	
	Gs	2.62	
Direct shear	Cohesion (C)	7 kPa	
	Friction angle	0	

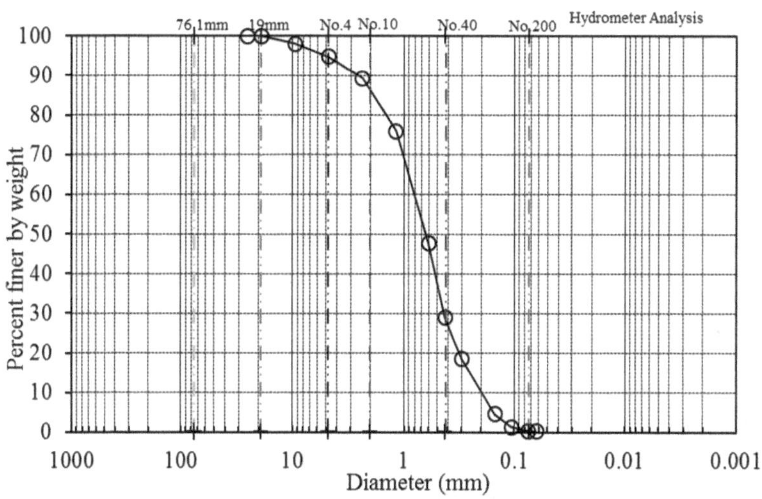

Fig. 1. Grain size distribution of sand

Table 2. Grain size distribution test results

Sieve size	25	19	12.50	9.5	4.75	2.36	1.18	0.6	0.425	0.3	0.150	0.106	0.075
Passing (%)	100.00	100.00	98.04	94.78	94.78	89.38	76.00	47.61	28.97	18.67	4.63	1.37	0.42
Effective diameter (D10) = 0.196					D30 (mm) = 0.433		D60 (mm) = 0.806			Uniformity coefficient (Cu) = 4.122			
Description: fine to coarse sand, trace of gravel, trace of silt										Coefficient of curvature(Cc) = 1.19			
%Gravel = 5.22				%Sand = 94.37				%Silt = 0.42			%Clay = 0		
Poorly graded sand													

Table 3. Properties of the Sand Used in the mix

Test type	Property	Value	Standard
Relative density	Maximum void ratio	68.48	Egyptian code for soil mechanics and design and executing the foundation 2009, part 2
	Minimum void ratio	47.35	
	Minimum density (gm/cm^3)	1.57	
	Maximum density (gm/cm^3)	1.80	
Compaction	Maximum dry density (KN/m^3)	18.02	
	Optimum moisture content (%)	9.87	

2.3 Cement Dust

Cement dust is a solid waste material produced from the cement industry. The cement dust used in this research comes from Torra factory in Egypt. Chemical analysis test was performed to determine the cement dust component as shown in Table 4.

Table 4. Chemical analysis results for cement dust

Component	Tio_2	Sio_2	Al_2o_3	Fe_2o_3	F-	Cao	K_2o	Na_2o	Mgo	Cl^-	So_3	L.O.I	Moisture
Percentage (%)	N.D	0.97	2.4	0.8	N.D	35	15	6	1.7	19	7.1	12	0.03

2.4 Lime

Lime material is used widely for the stabilization of soft clay. Chemical analysis test was performed on lime sample to know the component percentage of the lime sample as shown in Table 5.

Table 5. Chemical analysis results for lime

Component	Tio2	Sio2	Al2o3	Fe2o3	F-	Cao	K2o	Na2o	Mgo	Cl-	So3	L.O.I	Moisture
Percentage (%)	N.D	12.89	1.62	0.07	0.15	55.45	N.D	0.16	0.12	0.24	3.45	25.13	0.43

2.5 Encasement Material (Geotextile)

In this study, non-woven geotextile was used. Geotextile have a lot of advantages as it used for confining the column soil inside it, the geotextile conduct a good interface between the columns component and clay, so the friction produced between the soil and column is very high. The geotextile material is work as a filter as it allows the water to enter the column in order to consume it by cement dust to strengthen the column. Some tests have been performed on it to know the different properties of it. The tests performed are tensile strength, fabric weight, thickness and puncture and the results shown in the Table 6.

Table 6. Geotextile properties

Non-woven geotextile test			
Test	Test method	Result	Unit
Fabric weight	ASTM D-5261	438.51	g/m^2
Tensile strength	ASTM D-4595	5.86	KN/m
Thickness	ASTM D-5199	2.44	mm
Puncture	ASTM D-4833	260.24	N

3 Experimental Program

3.1 Mix Trials

Different percentages of cement dust, lime and sand were mixed to know the optimum percentage of each material in the mix. Nine samples were mixed and put into geotextile. The samples were submerged in water for 7 days. After 7 days, every sample were cut and tested to know the unconfined compressive strength of each sample. The sample that has the highest compressive strength is the optimum mix. Figure 2 shows the sample after prepared and submerged in water. The Table 7 showed the component percentage of each sample. After performing the tests on the samples, the tests showed that the optimum percentage of lime is between 4% and 6% and the optimum percentage of cement dust is from 25% to 30% as shown in the Fig. 3. According to this test, the cement dust, lime and sand percentage used are 30%, 4% and 66% respectively.

3.2 Physical Model Setup

A three dimensional steel model with dimension (1.4 m * 1.6 m * 0.7 m) was used to simulate the soft clay in laboratory at Construction Research Institute (CRI) – qualiobya government. The model is consists of steel tank with the previous dimension as shown in Fig. 4. Filter layer that consists of pipe network imbedded in sand filter layer was installed in the bottom of the tank to increase the rate of clay consolidation by making two paths for the water as shown in the Fig. 5. Figure 6 shows the pipe network details. Loading frame was installed. The loading frame consists of 2 supports

Fig. 2. Samples after preparation and submerging in water

Table 7. The sample component percentage

Sample no.	Lime %	Cement dust %	Sand %
1	4	20	76
2	4	25	71
3	4	30	66
4	4	35	61
5	4	15	81
6	10	30	60
7	8	30	62
8	6	30	64
9	2	30	68

(steel support), steel beam rested on the two supports and the third component is the loading instrument. The two supports are consists of steel angles painted with an isolated material. The frame was rested on the floor and tied with the tank with steel rod as shown in Fig. 7. The loading instrument is consists of body, rod to carry load, and arm that transfer the load as shown in Fig. 8. The load transferred 10 times bigger.

3.3 Kaolinit Preparation

The clay layer was prepared from kaolinit. The kaolinit is an artificial material used to simulate the clay behavior. To make a clay layer the kaolinit was mixed with water content that equal 150% from kaolinit liquid limit and well mixed with mixer for 15 min. After that the clay poured in the tank to form the clay layer (50 cm). Figure 9

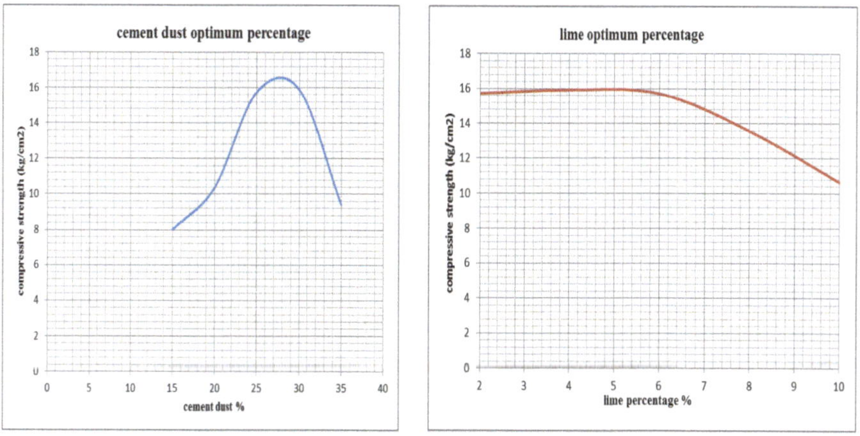

Fig. 3. Optimum percentage of cement dust and lime

Fig. 4. Physical model.

shows the tank after the formation of clay layer. After the pouring of soft clay, the consolidation process was begun by Appling concrete slab with dimension equal to tank dimension as shown in Fig. 10 to reach shear strength equal 7 kPa. The final thickness of the clay after reaching the required shear strength is equal 35 cm.

3.4 Column Formation and Installation

The geotextile used in this study is non-woven geotextile. This geotextile used as a filter in normal cases. The importance of geotextile used is to confine the column soil in

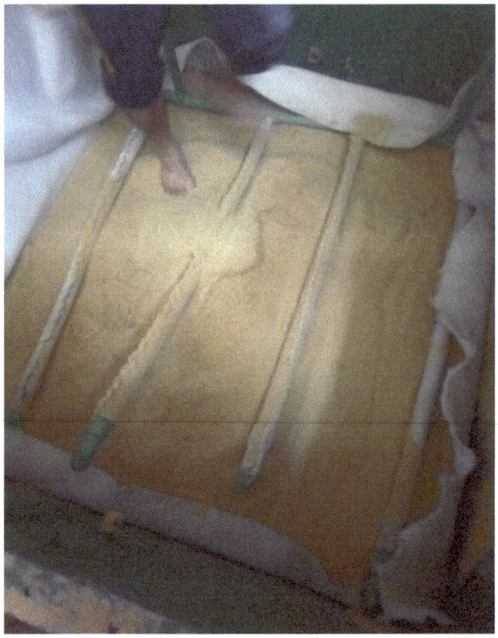

Fig. 5. Drainage system under the clay layer.

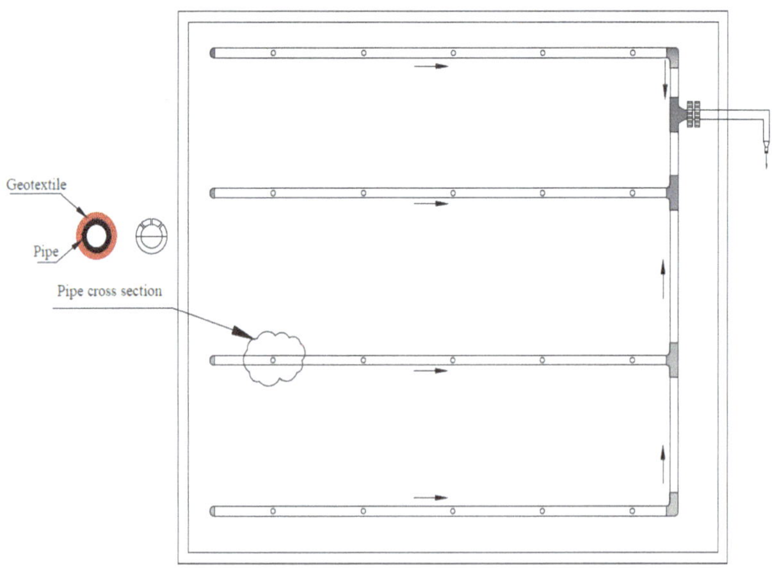

Fig. 6. Pipe network detalis

Fig. 7. Loading frame elements.

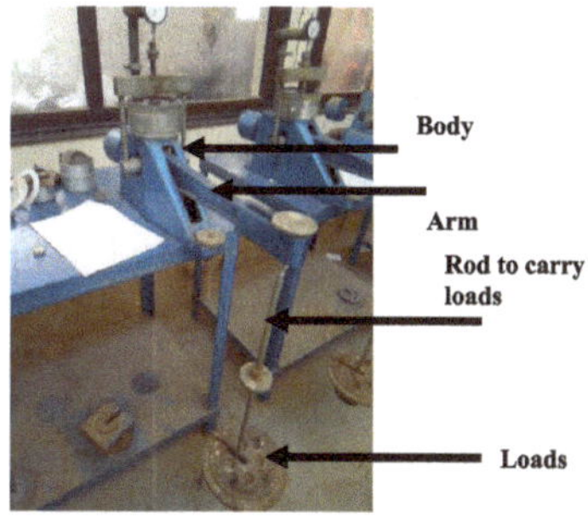

Fig. 8. Loading instrument component.

Fig. 9. Soft clay layer preparation.

Fig. 10. Consolidation process.

it to control the column in the clay and work as a filter to allow the water to enter the column soil inside it without any clay particles. The geotextile prevent the column soil from getting out and the geotextile increase the shear interface between soft clay and the column to increase the friction so the column can be work as a friction pile. To sew the geotextile to form the columns with the required diameters (3.5 cm, 4.0 cm and

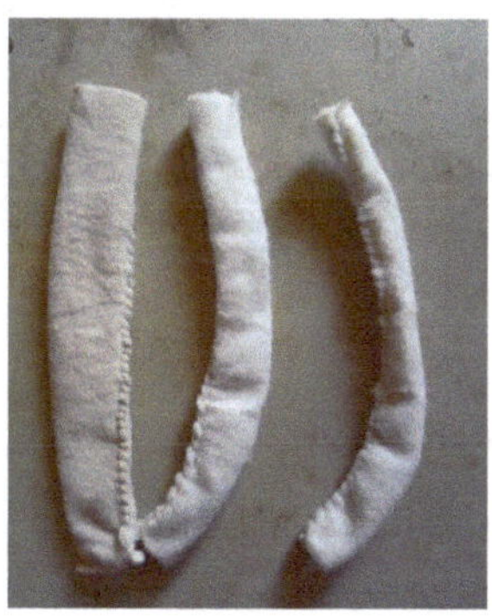

Fig. 11. Geotextile after preparation

5.0 cm), three plastic pipe were used with the required column diameter. The geotextile was wrapped around the pipe and sewed with overlap 5.0 cm to from the column outer cover as shown in Fig. 11.

After the clay layer, the geotextile, the mix and the geotextile prepared, columns installed in the clay layer with displacement method. The columns were prepared first and began to install in the clay in triangle shape as shown in Fig. 12. LVDT was installed to know settlement corresponding to each load as shown in Fig. 13.

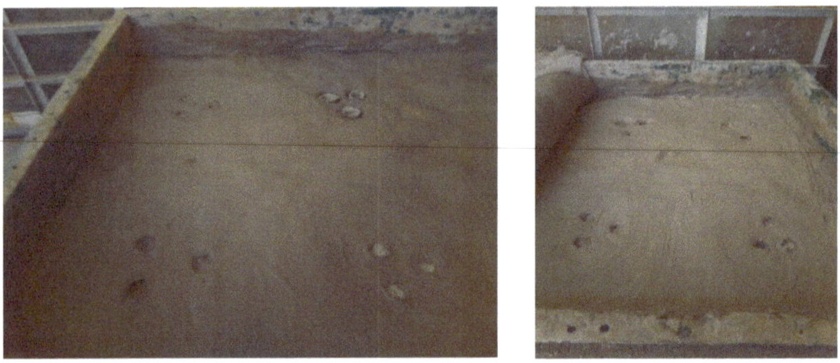

Fig. 12. Columns installation.

Fig. 13. LVDT installation.

3.5 Experimental Test Methodology

The main testing program consisted of total of 10 tests. Table 8 shows the experimental test configuration.

Table 8. Experimental test configuration

Test no.	Size of the footing	Column length	Column diameter	Spacing
T 0 E	Initial case			
T 1 E	15 cm	60% (21 cm)	3.5 cm	3D
T 2 E		80% (28 cm)	3.5 cm	
T 3 E		100% (35 cm)	3.5 cm	
T 4 E		60% (21 cm)	4.0 cm	
T 5 E		80% (28 cm)	4.0 cm	
T 6 E		100% (35 cm)	4.0 cm	
T 7 E		60% (21 cm)	5.0 cm	
T 8 E		80% (28 cm)	5.0 cm	
T 9 E		100% (35 cm)	5.0 cm	

4 Results and Discussions

The bearing capacity can be defined with the following methods

- Terzaghi method, where The bearing capacity is corresponding to the settlement equal 10% of the footing width (Yaseen et al. 2009; Zakariya 2001)
- De Beer (1967) method, where the bearing capacity can be taken at the break point between two straight lines with different slope.
- Hughes and Withers (1974) method, the bearing capacity is corresponding to settlement that equal 26 times of the undrained shear strength of the clay.
- Rao et al. (1997), the capacity is taken as the load corresponding to a settlement equal to 0.1 times the diameter of the stone column.

From the previous methods, the first method was chosen to be the method of determining the results

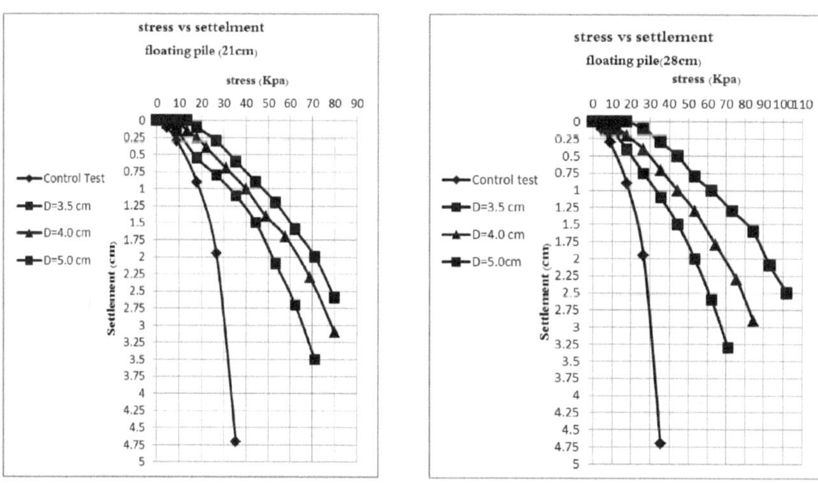

Fig. 14. Floating columns results (stress vs. settlement)

4.1 Effect of Cement Dust - Lime Floating Columns

Floating columns were installed and tested and the results were collected. The results showed that the floating columns improve the soft clay properties as the column soil consume the clay water content to be strengthen.so; a consolidation process was estimated to the clay that led to increasing in bearing capacity of the clay. After the column strengthen, the column behaves as a concrete pile because of the geotextile that produce a friction interface between the clay and the column that lead to increase in bearing capacity and reducing settlement. Figure 14 showed the results of the floating columns.

4.2 Effect of Cement Dust – Lime End –Bearing Columns

The results proved that the floating columns gives better improvement than end-bearing columns because of the low compressive strength of the column and the geotextile is weak against tensile strength, so the column was bulged. Figure 15 showed the results of end-bearing columns.

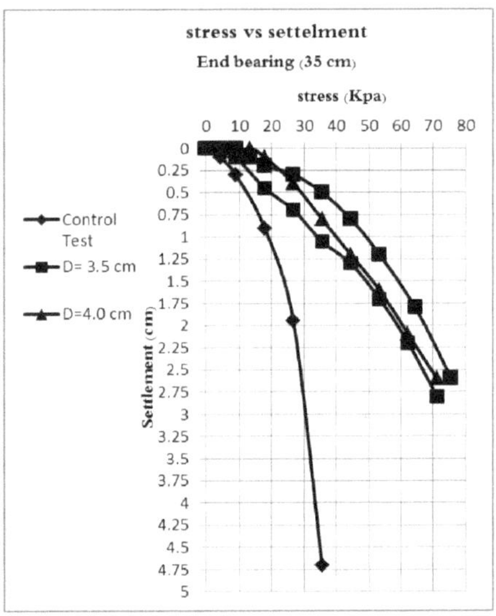

Fig. 15. End – bearing results (stress vs. settlement)

4.3 Effect of Increasing Column Diameter and Length

The results showed that the increasing in column length and diameter, the improvement percentage increased because the friction area increased and the soil consumes more water from clay water content, so the consolidation process rate increased. Figures 16 and 17 shows the effect of increasing column diameter and length on improvement in floating and end – bearing cases.

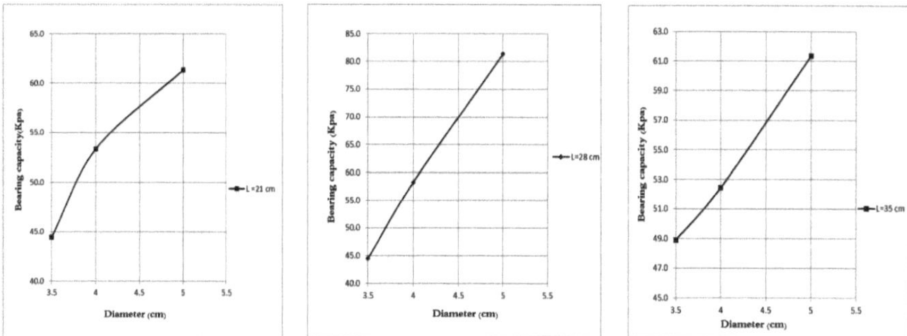

Fig. 16. Effect of increasing column diameter on bearing capacity

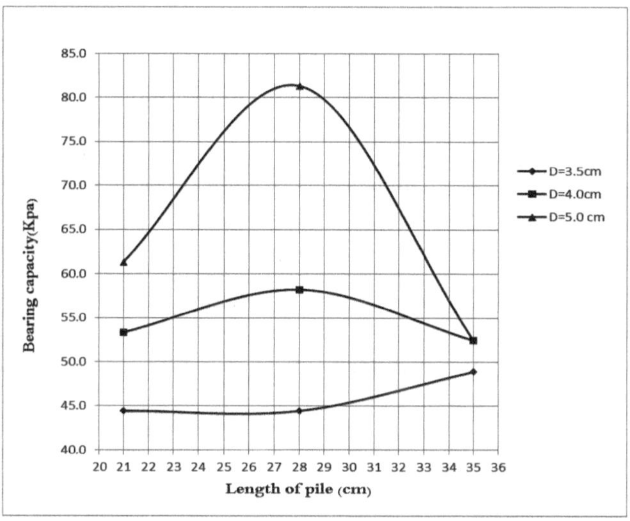

Fig. 17. Effect of increasing column length on bearing capacity

4.4 Bearing Capacity Improvement Percentage (I.P)

The bearing capacity improvement percentage was calculated from the following equation

$$I.P = \frac{Q_{treated}}{Q_{untreated}} * 100$$

Table 9 showed the Improvement percentage of all cases

Table 9. The bearing capacity improvement percentages

Case	Diameter (cm)	Length (cm)	Bearing capacity (Kpa)		Improvement (%)
			Untreated	Treated	
No column	–	–	22.89	–	–
With columns	3.5	21		44.4	194%
		28		44.4	194%
		35		48.85	213%
	4.0	21		51.9	227%
		28		57.74	252%
		35		51.075	223%
	5.0	21		59.98	262%
		28		80.7	353%
		35		58.85	257%

5 Conclusions

This study was performed to study the effect of adding cement dust – lime column in soft clay on the properties of soft clay. From the tests performed and the results collected, the following conclusions were drawn:-

1. The bearing capacity was increased and the settlement was decreased by adding the cement dust – lime column.
2. The floating columns give better improvement than end – bearing columns because of low compressive strength of the column and the flexibility of the geotextile.
3. The increase in column diameter and length lead to better improvement.
4. The maximum bearing capacity improvement percentage is equal 353% and the minimum bearing capacity improvement percentage is equal 194%.

References

ASTM D-5261. Standard Test Method for Measuring Mass per Unit Area of Geotextiles

ASTM D-4595. Standard Test Method for Tensile Properties of Geotextiles by the Wide-Width Strip Method1

ASTM D-5199. Standard Test Method for Measuring the Nominal Thickness of Geosynthetic

ASTM D-4833. Standard Test Method for Index Puncture Resistance of Geomembranes and Related Products

Azad, S.: Influence of soil type on stabilization with cement kiln dust. In: Construction and Building Materials, vol. 3 (2000)

De Beer, E.E.: Bearing capacity and settlement of shallow foundations on sand. In: Proceedings of Symposium on Bearing Capacity and Settlement of Foundations, Duke University, North Carolina, Lecture, vol. 3 (1967)

Al-Mosawe, M.J., Albusoda, B.S., Yaseen, A.S.: Bearing capacity of shallow footing on soft clay improved by compacted cement dust. J. Eng. **15**(4), 4417–4428 (2009)

Egyptian code for soil mechanics and design and executing the foundation, part 2 (2009)

Hughes, J.M.O., Withers, N.J.: Reinforcing of soft cohesive soils with stone columns. Gr. Eng. **7** (3), 42–49 (1974)

Khadhair, A.A.: Stabilization of Nasriyia sand dunes by cement dust. M.Sc. thesis, Buildings and Construction Department, College of Engineering, University of Technology, Baghdad (1997)

Fattah, M.Y., Al-Saidi, A.A., Jaber, M.M.: Improvement of bearing capacity of footing on soft clay grouted with lime-silica fume mix. Geomech. Eng. **8**(1), 113–132 (2015)

McCoy, W.J., Kriner, R.W.: Use of Waste Kiln Dust for Soil Consolidation. Lehigh Portland Cement Co., Allentown (1971)

Rao, S.N., Reddy, K.M., Kumar, P.H.: Studies on group of stone columns in soft clay. Geotech. Eng. J. Southeast. Asian Geotech. Soc. **28**, 165–181 (1997)

Zakariya, W.A.S.: Soil improvement using stabilized and non-stabilized stone columns with different reinforcement configuration. Ph.D. thesis, Building and Construction Engineering Department, University of Technology, Iraq, p. 142 (2001)

Best Practice and Minimum Standard in Prefabricated Vertical Drain Execution

Robert Thurner[1,2(✉)] and Mohamed Ayeldeen[2]

[1] Keller Grundbau GmbH, Vienna, Austria
r.thurner@kellerme.com
[2] Keller Grundbau GmbH, Dubai, UAE

Abstract. Even if the basic idea of applying Prefabricated Vertical Drains in combination with surcharge to improve soft soils seems very common and to be a simple process there are many details and basics to be considered to deliver a proper product and achieve expected results at the end with a safe operation. This paper concentrate on relevant practical points during the initial phase and the main execution of such projects. The included topics are ranging from material, equipment, installation and monitoring. One of the most important topics is about safety to ensure a safe production reaching the goal of zero accidents. Since the basic idea is a combined application of PVD with a surcharge both parts of the system has to be executed in the correct way and monitored accurately, mainly using settlement points.

1 Introduction

Several types of problematic soils are distributed all over the world which need to be improved before starting any further engineering activities. One of the most common problematic soils worldwide is soft soil which is recognized with low shear strength, high compressibility and low bearing capacity. Several techniques have been applied successfully to improve the soft soil such as mixing with binders or fibers or even a combination between binders and fibers (Ayeldeen and Kitazume 2017). Stone column as well shows noticeable success as improvement technique for soft soils (Sivakumar et al. 2007), however one of the most economics and ecofriendly techniques is to pre-consolidate the soft soil which will eventually increase its stiffness and reduce the expected settlement. The pre-consolidation process can be performed by using preloading pressure (surcharge) in combination with or without vertical drains in the soil. Generally, the idea behind the Prefabricated Vertical Drains is to shorten the drainage path of the water from clay layers by installing vertical drains which allow the groundwater to migrate inside the clay layer to drain horizontally towards the vertical drains instead of taking a longer drainage path equals to half the thickness of the clay layer or it might be the whole clay thickness if the drainage is allowed from one side only (Indraratna et al. 2015). In other words, it accelerates the consolidation process during the pre-loading which will lead to an increase in the soil stiffness and reduction

H. Shehata and H. Poulos (Eds.): GeoMEast 2018, SUCI, pp. 66–76, 2019.
https://doi.org/10.1007/978-3-030-01917-4_5

in the final settlements for the final/long term stage. Soil stabilization with PVD's is used on compressible and saturated soils. These soils are characterized by their soft structure, big pore capacity normally filled with water (fully saturated). Exposing the soft clay soil to high value of surface stress - such as an embankment or a building - a remarkable settlement can be noticed due to the high compressibility of such soil. This settlement could create serious construction problems in short or long term. The stress created by the surcharge is initially carried by the pore water. Due to the low soil permeability, excess pore water pressure will decrease gradually with increasing the effective stress at the same time. However this process shall take long time due to the low permeability of the soft clay.

PVD's consist of a prefabricated strip which is very suitable for water transportation. The flexible core is typically manufactured of polypropylene. Typically, both sides have grooves, through which water can flow unimpeded. The core is wrapped in a strong and durable geotextile filter fabric with excellent filtration properties, allowing free access of pore water into the drain. At the same time, this filter prevents piping of fines from adjacent soils without clogging.

2 Material and Execution

The system of preloading soft soil with vertical drains is a combined system between PVD installation process and preloading process later on. The used materials in PVD stage are the vertical drains itself and the anchor elements used to keep the drains in its location after drilling down.

2.1 Drain Material

The Prefabricated Vertical Drains (PVD) are composed on a geotextile filter jacket surrounding a tough, strong drain core. The PVD material is responsible for the drainage of water from the clay layer accelerating the consolidation process.

The PVD material should be tested to confirm the standards. The following tests are the minimum testing to be conducted on PVD samples:

Tensile properties of Geotextile	ASTM D4595-17
Discharge Capacity	ASTM D4716
Determining Apparent Opening Size	ASTM D4751-16
Water Permeability of Geotextile by Permittivity	ASTM D4491-17

2.2 Anchor Elements

Anchor plate/anchor rebar – steel elements attached to the bottom of the PVD holds it in place in the soil as the mandrel is extracted. Anchor plate and rebar should fit to the mandrel tip to avoid getting soil inside the mandrel. As shown in Figs. 1 and 2.

Fig. 1. Anchor plate

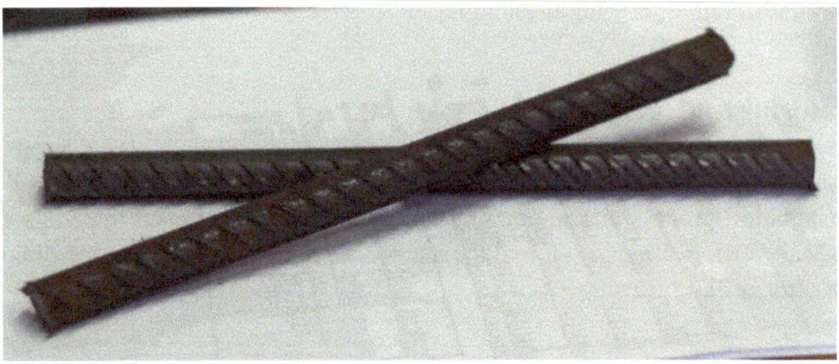

Fig. 2. Anchor bars as an alternative to anchor plates #8–10 mm L = 20 cm.

Anchor plates	Rebars
Advantages: • workability • closing of the open tip of the mandrel against penetration of soft soil – less cleaning work	Advantages: • easy to manufacture on the site • lower price • can be used as a backup/alternative
Disadvantages • risk of weak welds • risk of cutting PVD's by the sharp edges of anchor plate • higher price than rebars • the site depends on the supplier (risk of delay deliver) • has to comply with the specs	Disadvantages • workability • Affecting the production rate as percentage of laying down the masts are increasing because of improper closing of the steel bars

2.3 Working Platform

The working platform is the upper layer above the virgin soil which will be used as a platform for the installation and later it will work as upper drainage layer for the excess pore pressure. Working platform should be levelled, compacted and suitable for traffic of heavy machines in all weather conditions with minimum thickness of 1.0 m. The used material should guarantee both the proper bearing capacity and high permeability, therefore the following properties can be sufficient for the used cohesion less soils:

- Friction angle min $\varphi = 30°$
- Ability to compact
- High permeability (min 10–3 m/s)
- Long service life (The working platform layer should be suitable for traffic of all types of machines in all weather conditions)
- The fines content for the drainage layer should be less than 10% based on the experience and technical point of view

In some cases a bearing steel plates on top of this platform can be used as additional platform to increase the rig stability which can be needed in very soft soil conditions. This additional platform is also helpful during laying the mast down for maintenance purposes where the contacting stress will highly increase in the front part of the rig due to increasing the eccentricity with changing the mast to be horizontal as shown in Fig. 3. Geotextile can be used below this cohesion less material as well in case of large soft spots considering minimum of 50 cm overlap to be placed. Geotextile must be pinned to the ground every 2 m with U/T-bar. Sand laying must be supervised to avoid geotextile destruction with minimum thickness of 50 cm as shown in Fig. 4.

In general an installation of a thin geotextile (approx. 100 g/m^2) below the working platform on top of the soft soil is recommended to limit the material consumption and to increase the stability & durability for the platform.

Fig. 3. Steel plates (1500 × 1500 × 20 mm) used during maintenance and soft spots

Fig. 4. Geotextile sheets for the large soft spots

2.4 Execution (Planning and Production)

Vertical drain execution process is long term process of repetitive work which need accurate planning to increase the productivity and the accuracy. A full checklist should include the following items:

- Handing over a working platform fulfilling all technical requirements (thickness and material)
- Levelling and compacting
- Checking the stability using a trial dummy excavator or any other reliable method to check the working platform
- Setting out the drains locations by plastic sticks (recommended), tapes or anchor plates
- PVD execution with respect to all safety rules:
  - Material transport and traffic management on site
  - Equipment certifications, maintenance and repair
  - Handling of material and cutting during the installation process

Figure 5 is presenting production progress on PVD site. Nylon tires or rubber sheets can be used as a soil scrapper to prevent the soil from entering inside the mast during pulling the mandrel up as shown in Fig. 6.

3 Monitoring the Performance

After the PVDs are installed, the area should be loaded with a specific height for the filling material to apply the required stress according to the design to start the consolidation process. Before and during the preloading period, the settlement should be monitored periodically with a sufficient accuracy (typically within ±2 mm for the nivellement). In particular for bigger areas it could be required to install intermediate fixed points, e.g. sort of micro piles to ensure a proper accuracy.

The frequency of such measurement is depending on the phase you are in. During the installation and the beginning of the preloading measurements should be taken at least twice a week, whereas later on bi-weekly could be sufficient.

Fig. 5. Production progress and planning on PVD site

Fig. 6. Rubber sheets to seal the mast groove

During the installation of the surcharge unfortunately the settlement plates are damaged or destroyed partially from the earth working process if not done properly. Therefore, a certain robustness and redundancy should be incorporated in the measurement system by sufficient (spare) points and some protection measures.

3.1 Instrumentation

For major projects the execution of a well monitored trial field is recommended. This trial field should be executed in advance, to use the received knowledge for the final design and layout of the major works. To bridge the malfunction of some installed devices later on a redundant systems should be developed using the different tools available. A specialized engineer having previous experience in the installation and monitoring of the instrumentation shall be responsible of the monitoring system, which shall include the following:

Piezometers
The piezometer is a device that is used to measure the pore water pressure. It can be installed at any depth (specified by the design engineer). Monitoring the pore pressure readings shall help in understanding the behaviour of the soil layers during consolidation process. Also, it can be used to measure the degree of consolidation with time form the pore water pressure measurements. Installing the piezometer within the area of the installed PVD leads to an uncertainty for the interpretation of the results, since the distance of the device and the PVD cannot be controlled or measured. In addition the installation of the piezometer should be done prior to the PVD to understand the pore water pressure distribution. The data should be compared with CPTu data and the measured water table in a trial pit.

Interpretation of those data might be the most difficult task for the trial field evaluation. This is valid in particular for the layers where the PVD are installed, since the distance between the single nearest PVD and individual measurement device cannot be exactly be defined.

Extensometer
The extensometer is a settlement measuring device and for soft soils spider type extensometer are mandatory with telescopic parts to avoid buckling of the extensometer tubes. It consists of magnets that are positioned along the length of the pipe. These magnets are called spider magnets and they are coupled to the soil and free to move up or down with the soil. This device shall give a proper view for the settlement of each layer during the preloading period, which is of particular interest for cases where the PVD do not cover the full depth of soft soils. In such cases the settlement contributions from the treated and untreated layer can be distinguished and used for the evaluation of the applied models during design.

Inclinometer
It measures the lateral movement and horizontal deformities of the soil. For the inclinometer a clear orientation on top should be ensured as well as they should be installed without a torsion to avoid change of orientation by depth. Possible problems with the stability of the slope during fill placement can be determined using the inclinometer.

Settlement Points
Steel rods with steel or RC base which are placed after the installation of PVD should give the overview about the surface settlements. The settlement point shall be extended above the top of the surcharge layer to allow the surveyors to monitor it after placing the full height of the surcharge. A detailed cross section for the typical settlement point is presenting in Fig. 7.

It is essential to ensure the protection of all the installed instrumentation during applying the surcharge. Therefore, all the above mentioned devices (excluding settlement points) must be protected using steel protection towers as shown in Fig. 8.

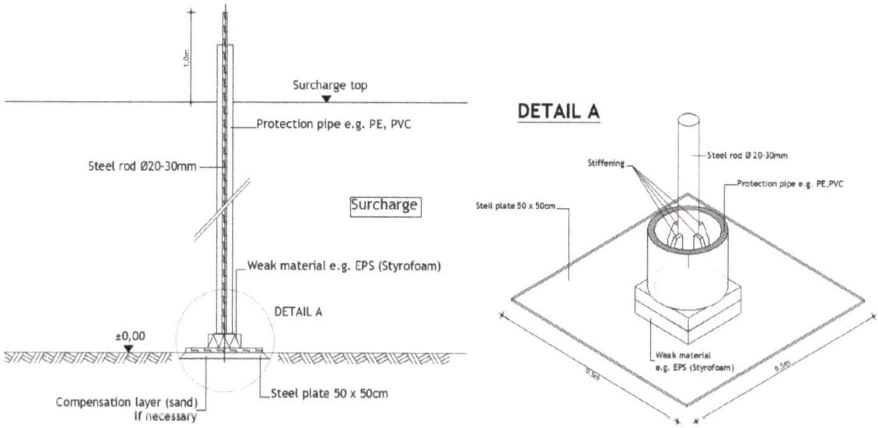

Fig. 7. Typical settlement point

Fig. 8. Steel protection tower for instrumentation protection during preloading

An area should be chosen to be the project trial zone in which the instrumentations will be installed, a sufficient number of settlement points, trial pits and boreholes to monitor the Ground Water Table (GWT). The trial zone is the prototype for the rest of the project, the acceptance criteria shall be based on the results obtained from the trial zone monitoring including possible variation of the soil throughout the site. Pre and Post preloading CPTu can be performed in the trial zone to have a clear view of the change in the soil parameters with time if any. Dissipation tests (excess PWP $\ll$ 50%) have to be performed to evaluate the coefficient of horizontal consolidation which is one of the key parameters in the design of the PVD. The results of the devices are reviewed and analyzed by an experienced geotechnical engineer during the whole monitoring period. The engineer shall be responsible of analyzing the results and to make sure the devices are giving reasonable results during each stage.

Some calculations can be preformed to estimate the degree of consolidation, DOC, from the piezometer reading (Indraratna et al. 2012, Bergado et al. 2002) and to be compared with the calculated degree of consolidation from the settlement (Asaoka 1978), however both values could be incomparable in some cases as the estimated DOC from piezometer is calculated at a certain point not as the average DOC which is calculated from the measured settlement (Chu et al. 2005). The degree of consolidation can be evaluated from the results of monitoring devices, and it can be used to calibrate the soil parameters used in the design and estimating the future settlement for the whole project.

4 Preloading Surcharge

The surcharge unit weight and height must - in combination - satisfy the required stress on the platform level. Field unit weight test (sand cone test) should be performed to calculate the soil unit weight and compare it to the design requirements. Establishing and sequence of surcharge construction have to be done according to the contract and design specifications. The quality of the used material and the efficiency of compaction (mainly done by the transporting equipment) has to be approved by the contractor. The final height of the overload embankment will be specified after check of the unit weight. The height of surcharge will be adjusted to obtain the load specified in the design. The placing and compaction of the permanent fill materials and the material itself has to fulfill the project and design requirements with respect to the foundations built thereon. Furthermore, a site-specific correlation between CPTu and density of the surcharge can be established to ensure during the evaluation of the system surcharge – PVD - soil, the correct stress is applied.

To ensure the proper function of the installed PVD the surcharge should be placed shortly after the installation of the PVD to avoid any reduce effectiveness e.g. due to possible clogging of the filter material.

5 Safety During Execution

According to what has been mentioned previously that PVD execution process is a fast repetitive work, therefore high level of safety should is essential on site all the time. Several hazards can threaten the safety of execution such as:

- Inadequate working platform and its stability
- Equipment/machinery quality and its reliability
- Inexperienced people
- Lack or poor communication between people
- Movement of machinery and work management
- Improper organizational structure and fuzzy responsibility
- Inexperienced or incorrectly selected subcontractors
- Housekeeping/waste management
- Site layout and organization of construction facilities
- Welding works
- Lifting operations and movement of the materials
- Lighting
- Manual handling
- Drugs/alcohol
- Weather conditions

For All and Each from the Previous Points Has to Be Studied and Checked Regularly by the Safety and Project Teams According to the International Standards and Rules

5.1 Safety Rules

To mention one of the rules for example, the safe zone of one rig should be at least 40 meters around it (depending on the height of the rig). Every rig must be outside this safe zone of another rig as presented in Figs. 9 and 10 for the wrong and right locations.

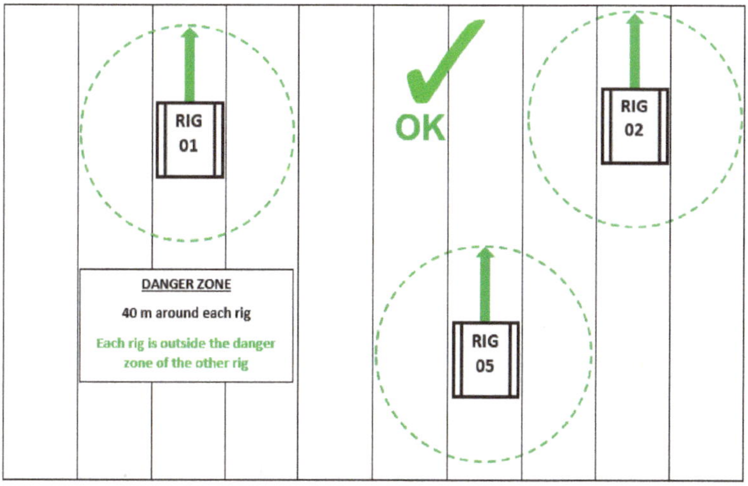

Fig. 9. Correct placement of rigs on the working platform

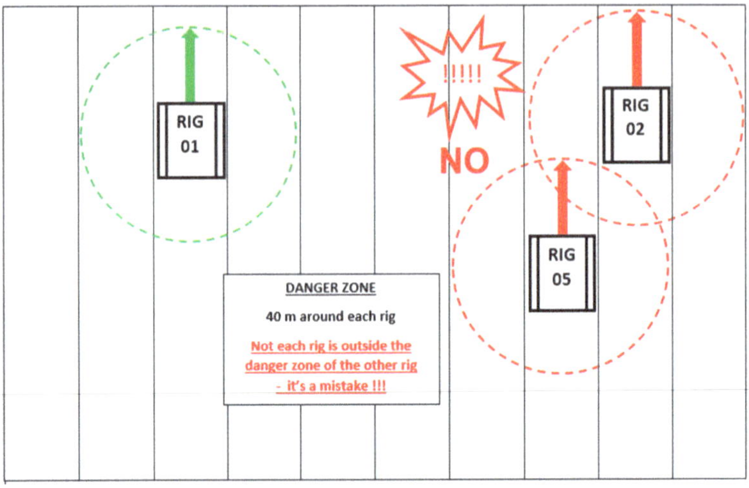

Fig. 10. Incorrect placement of rigs on the working platform

6 Conclusions

To ensure a technical and safety wise successful execution of a PVD project the entire system consisting of PVD, surcharge and monitoring has to be planned and execute following some minimum standards and applying best practice. Some basics have been described and summarized in this paper since unfortunately on several sites executed in the recent past some points have been missed and caused a negative impact to the client.

References

Asaoka, A.: Observational procedure of settlement prediction. Soils Found. **18**(4), 87–101 (1978)

Ayeldeen, M., Kitazume, M.: Using fiber and liquid polymer to improve the behaviour of cement-stabilized soft clay. Geotext. Geomembr. **45**(6), 592–602 (2017)

Bergado, D.T., Balasubramaniam, A.S., Fannin, R.J., Holtz, R.D.: Prefabricated vertical drains (PVDs) in soft Bangkok clay: a case study of the new Bangkok International Airport project. Can. Geotech. J. **39**(2), 304–315 (2002)

Chu, J., Yan, S.W.: Application of the vacuum preloading method in soil improvement projects. In: Elsevier Geo-Engineering Book Series, vol. 3, pp. 91–117. Elsevier, Amsterdam (2005)

Indraratna, B., Chu, J., Rujikiatkamjorn, C.: Ground Improvement Case Histories: Embankments with Special Reference to Consolidation and Other Physical Methods. Butterworth-Heinemann, Oxford (2015)

Indraratna, B., Rujikiatkamjorn, C., Balasubramaniam, A.S., McIntosh, G.: Soft ground improvement via vertical drains and vacuum assisted preloading. Geotext. Geomembr. **30**, 16–23 (2012)

Sivakumar, V., Glynn, D., Black, J., McNeill, J.: A laboratory model study of the performance of vibrated stone columns in soft clay. In: Proceedings of the 17th International Conference on Soil Mechanics and Foundation Engineering, Madrid, Spain, vol. 3, pp. 1545–1350 (2007)

Efficacy of Sustainable Soil Stabiliser on Compaction Properties of Expansive Soil

Asita Dalvi(✉), Abhilash Patil, Yashodip Jadhav, Khandu Jadhav, Pritesh Nehatrao, Sanjay Gawali, and Shardul Mantri

Sinhgad Institute of Technology and Science, Pune, India
asitadalvi@gmail.com

Abstract. New advances in soil stabilization technology are increasingly being used in both constructing and improving/rehabilitating unsurfaced and paved roads worldwide. With rapid industrialization and urbanization, the non-useful waste material is creating human as well as environment concern. Thus making it necessary to address the issue of reuse of such non-useful waste materials to provide a sustainable solutions and giving rise to the development and introduction of other kinds of soil additives such as plastics, bamboo, liquid enzyme soil stabilizers, providing Geosynthetics, etc.

The present thus study deals with experimental investigation carried out to verify the efficacy of various soil stabilizers on expansive soil from Baramati, Maharashtra. The soil stabilizers ranging from traditional to recent advancements such as; lime, plastic bottle chips (1 cm × 1 cm), Rice Husk Ash (RHA) in varied percentages are used. Also the suitability of providing geotextiles layer has been verified. Based on free swell index tests, the soil is confirmed to have very high swelling potential. A preliminary experimental investigation has been carried out by conducting standard proctor tests with and without stabilizer to determine compaction properties.

It has been observed that Rice Husk Ash and lime significantly affects the compaction properties. Plastic and geotextile although increases MDD, the OMC value does not change, suggesting further investigation. Interestingly, mixture of two additives does not affect the compaction properties, more combinations might be investigated. Finally, it is recommended that addition of RHA certainly helps providing a sustainable and economical solution in soil stabilization as presented in this work. It is well understood that the soil is invariably a nonlinear material; hence the results and conclusions may be soil specific, thus it is anticipated that present work will serve as a preliminary guideline to select suitable type of stabiliser for a particular soil for reliable results.

Keywords: Soil stabilization · Sustainable materials · Rice husk ash Geotextile

H. Shehata and H. Poulos (Eds.): GeoMEast 2018, SUCI, pp. 77–84, 2019.
https://doi.org/10.1007/978-3-030-01917-4_6

1 Introduction

In design and construction of any structure, the role of soil is very crucial. Since the soil is in direct contact with the structure, it acts as a medium of load transfer. Any structure need a stable foundation for their proper construction and life long durability. For a stable foundation, the underlying soil strata must be stable. Structures constructed on weak and unstable soil strata are more likely subjected to failures. To future problems in weak soil, soil stabilization is must.

Soil stabilisation means the improvement of stability or bearing power of the soil by the use of controlled compaction, proportioning and/or the addition of suitable admixture or stabilisers. Following the pace of modernization the materials and methodologies adopted for soil modification/stabilization also have experienced a transformation from agriculture and domestic waste to highly equipped machinery and synthesized stabilizers, to meet the growing requirements (Jayanthi 2016).

Due to rapid growth in population, industrialization and urbanization is increasing leading to the growth in the waste materials which should be handled and treated such that it can contribute in sustainable solutions to different problems. There are several methods used for soil improvement such as mechanical compaction, dynamic compaction, vibro-floatation, preloading, sand and stone columns, injection of special grouts, etc. along with inclusion of different stabilizers and admixtures.

In present study experimental investigation is carried out to verify the efficacy of various soil stabilizers on expansive soil from Baramati, Maharashtra. The soil stabilizers ranging from traditional to recent advancements such as lime, plastic bottle chips (1 cm × 1 cm), Rice Husk Ash (RHA) are selected and compared based on their compaction properties. The core perception of present work is to address the issue of reuse of such non useful waste materials to provide a sustainable solution and giving rise to the development and introduction of other kinds of soil additives such as plastics, bamboo, liquid enzyme soil stabilizers, providing Geosynthetics, etc.

It is known that many researchers like Maher and Ho (1994), Lindh and Eriksson (1990), Rao and Dutta (2004), Ameta and Wayal (2008), Bale (2011), Ashraf et al. (2011), Tiwari and Sharma (2013), Bhattarai et al. (2013), Negi et al. (2013), Ghosh and Ramesh (2014), Rajput et al. (2015), Gupta and Ameta (2016), Kumar et al. (2016), Rathan Raj (2016), have conducted investigation on stabilization of soils using different types of admixtures. However, the literature review presented herein is not limited but confined to sustainable materials used across the world to achieve soil stabilization as discussed in next section.

2 Literature Review

Kumar (2016) and Ashraf (2011) studied the use of plastic (LDPE) waste strip with dune sand for design of flexible pavement in construction of road and raw plastic bottles as stabilizers respectively. The results of tests demonstrated that inclusion of plastic waste in appropriate amounts increases strength and deformation behaviour of soil. Khan (2016) and Ghosh (2014) did the work on use of jute geo-textile for improving soil characteristics. Shear strength, dry density, permeability and CBR have been compared before and after lying of the jute geo-textile. While shear strength, dry density and CBR increased, permeability and penetration (check for settlement) decreased on introduction of jute geo-textiles, indicating significant improvement in the engineering behavior. Rathan Raj (2016) did an analysis to study the performance of Rice Husk Ash (RHA) as soil stabilizer. The effect of rice husk ash in alluvial soil and clay soil on the variation of index properties, compaction characteristics, shear strength and CBR values were analysed. Rajput (2015) did a study on effects on geotechnical characteristics of black cotton soil when fly ash is added to the soil. The test result showed a significant improvement in compaction and CBR characteristics. Fly-ash is found to be an effective waste material for the stabilization of expansive soil. Negi (2013) studied the complete analysis of the improvement of soil properties and its stabilization by using lime. Jayanthi and Singh (2016) studied the utilization of sustainable materials for improving certain defined properties such as plasticity, swell potential, strength, and density of the soil mass. In this paper, literature has been presented showcasing the superiority of the sustainable materials over the conventionally used soil stabilizers and the need for conducting further research to make these materials an easy and choicest replacement over the former.

However, it is observed that various soil stabilizers are available, however each of these are investigated independently, thus providing a scope to investigate the efficacy of various stabilizers on the same soil sample to evaluate the performance of each of them and the response of soil too. Thus, methodology and testing program has been decided as discussed in next section.

3 Methodology

Present work thus serves as a guideline to pick up appropriate and optimistic quantum of stabilizer for a specific soil. It is believed that with number of emerging stabilizers, there is need of a preliminary test to reduce time and achieve economy as well. As soil is non linear its response to a specific stabilizer will vary and thus such an attempt to compare test results based on compaction properties (as it forms the basis of all further tests such a CBR, swelling pressure etc.) and then selecting the best out of them and carry out further investigation is the core perception of this work.

In the present work, Index and Engineering properties of the soil sample such as Atterberg's consistency limits, free swell index and standard proctor tests are performed. According to the available literature, before performing any test on soil such as CBR test, plate load test, swelling pressure test etc. it is important to know its compaction properties. It is anticipated that present work will serve as a preliminary guideline to select suitable type of stabiliser for a particular soil for reliable results. Hence, to compare the compaction properties of soil with different percentage of stabilizers and admixtures, standard proctor tests are performed.

Standard proctor test is performed according to IS 2720-PART VII-1980 (Reaffirmed-2011). A cylindrical metal mould of size 100 mm diameter $\times$ 127 mm height along with a rammer weighing 2.6 kg is used. The soil sample is sieved through

4.75 mm sieve. The compaction is carried in 3 equal layers with 25 no. of blows for each layer.

According to the literature studied, soil stabilizers such as; lime, plastic bottle chips (1 cm × 1 cm), Rice Husk Ash (RHA), geotextiles(GT110), Natural Fiber (3 cm) in varied percentage are used. Figure 1 shows the soil sample with addition of plastic bottle chips. Also, a compaction test with a combination of two stabilizers has been investigated to study the effect of mixing 2 additives in soil stabilization. Table 1 summarizes the percentage of stabilizers used in present work. Thus, standard proctor tests as listed in Table 1 have been carried out and the results and interpretations are as given in subsequent section.

Fig. 1. Tray of plastic waste strips mixed with soil sample

Table 1. Type of stabilizer and its percentage

Admixture/stabilizer					
1	2	3	4	5	6
Plastic	Woven geotextile	Natural fiber	Lime	RHA	RHA + Plastic
0.25%	1 layer	0.25%	2.5%	0.5%	0.5% + 0.5%
0.5%	2 layer	0.5%	5%	1%	
1%			7.5%	2.5%	
2.5%			10%	5%	
5%					
7.5%					

4 Results and Summary

The prime objective of present work is to evaluate the performance of various soil stabilizers on expansive soil. Accordingly, the gradation of soil is found, then the swelling potential is determined and compaction properties are as well found. The results of index and engineering properties of soil sample are presented in Table 2. From Table 2, the soil is classified as CH with very high swelling potential.

The testing programme as outlined in Table 1 has been followed to summarize: the plastic strips are added ranging from 0.25% to 5%, lime is added from 2% to 10%, geotextile is added as 1 layer to 2 layers, natural fiber is added from 0.25% to 0.5% & RHA is added from 0.5 to 5%. Figure 2 shows the compaction graph comparing soil samples with varying percentage of RHA. From Fig. 2 the maximum value of MDD was found to 1.475 kN/m^3 on addition of 1% of RHA & at 25% water content. The highest recorded OMC and MDD values for each stabilizer is summarized in Table 3

Table 2. Soil properties.

Parameter	Results
Liquid Limit	68%
Plastic Limit	38%
Plasticity Index	30%
Free Swell Index	87.5%
Maximum Dry Density (MDD)	1.301 g/cc
Optimum Moisture Content (OMC)	35%

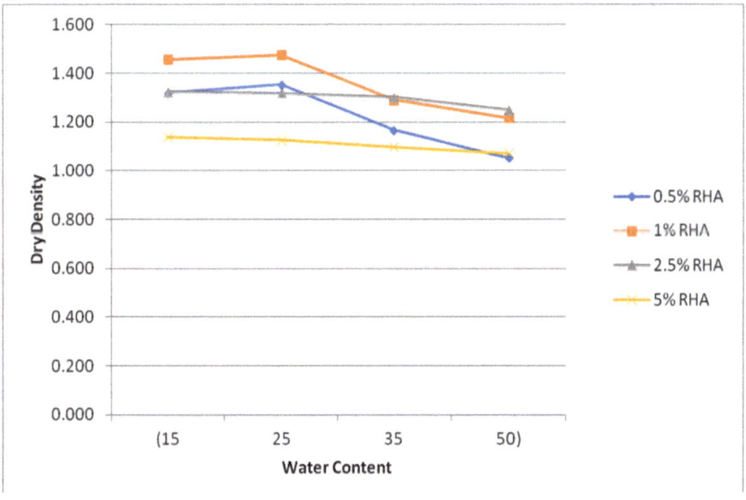

Fig. 2. Comparison of soil samples with varying percentage of RHA

Table 3. Summary of highest recorded OMC and MDD.

Sr. No	ADMIXTUE	OMC %	MDD g/cc	% Increase	% Change in OMC
1	SOIL	35	1.301	NA	-
2	5% PLASTIC	25	1.416	8.83	10% decrease
3	2 LAYER OF GEOTEXTILE	35	1.359	4.45	Unaltered
4	0.25% NATURAL FIBER	25	1.372	5.45	10% decrease
5	2.5% lime	25	1.492	14.68	10% decrease
6	1% RHA	25	1.457	13.37	10% decrease
7	0.5% RHA + 0.5%PLASTIC	25	1.325	1.84	10% decrease

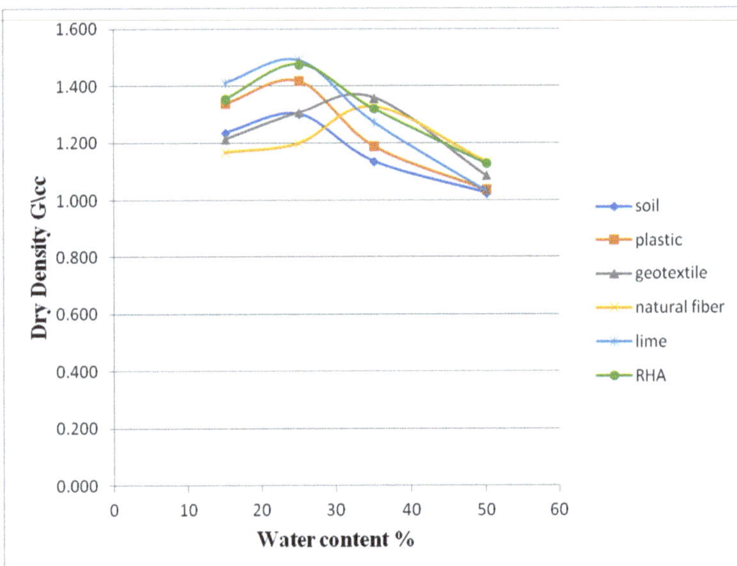

Fig. 3. Compaction curves for soil with different stabilizers

and also depicted in Fig. 3. From Fig. 3, It is clearly observed that the performance of soil with RHA and lime as stabilizers is good as compared to other stabilizers.

It can be clearly observed from Table 3, that there is significant increase in MDD with use of single stabilizer whereas addition of multiple stabilizers does not affect MDD value.

5 Conclusion

The present work is thus primarily based on experimental investigation on expansive soil and effect of various stabilizers in varied percentages with emphasis on compaction properties. The major conclusions drawn in present work are:

1. Rice Husk Ash and lime significantly affects compaction properties as it reduces OMC and increases MDD, yet it is to be noted that the % of RHA mixed is less as compared to lime. Furthermore, RHA proves to be a sustainable solution being a waste material.
2. Further investigation can thus be carried out w. r. t swelling pressure, CBR etc. by selecting the additive that yields optimistic results.
3. Similarly, addition of Plastic affect the compaction properties by reducing OMC & increasing MDD, however it is to be noted that addition of plastic might lead to genenvironmental concerns and thus not recommended.
4. Addition of Geotextile indicates increase in MDD; the OMC value does not change, suggesting further investigation.
5. Interestingly, mixture of two additives does not affect the compaction properties, hence more combinations might be investigated.

Thus, it is recommended that addition of RHA certainly helps providing a sustainable and economical solution in soil stabilization as presented in this work. The present work also ensures that it is important to select appropriate type and quantity of stabiliser to be used for better performance. Although, in case of expansive soil, improvement in swelling potential is a major part of stabilisation, further studies are being carried out to analyse the change in swelling potential with change in compaction properties of soil. Finally, the soil is invariably a nonlinear material; hence the results and conclusions may be soil specific, yet the present work is thus suggestive of a reliable solution as it helps to select a stabiliser in an optimistic way.

References

Khan, A.J.: Road subgrade reinforcement using jute geotextiles. Studied in Bangladesh, researchgate.net/publication/270881873 (2016)

Gupta, A., Ameta, N.K.: Fine sand stabilization using sanitary ware waste as admixture for design of flexible pavement in construction of roads. AJER **5**(8), 186–191 (2016)

Bale, A.S.: Potential reuse of plastic waste in road construction: a review. Int. J. Adv. Eng. Technol. (IJAET) **2**, 233–236 (2011)

Ashraf, A.: Soil stabilization by using raw plastic bottles. In: Indian Geotechnical Conference, Kochi, December 15–17, 2011 (Paper No. H-304) (2011)

Negi, A.S.: Analysis of the improvement of soil properties and its stabilization by using lime. Int. J. Innov. Res. Sci. Eng. Technol. **2**(2) (2013)

Ghosh, B., Ramesh, V.: Improvement of soil characteristics using jute geo-textile. Int. J. Sci. Eng. Technol. Res. (IJSETR), **3**(7) (2014)

Ameta, N.K., Wayal, A.S.: Effect of bentonite on permeability of dune sand. EJGE, Vol. 13-Bund. A (2008)

IS Codes- IS: 878-1956, IS 2720 (PART III, IV, VII) (1980)

Lindh, E., Eriksson, L.: Sand reinforced with plastic fibres, a field experiment. In: McGown, A., Yeo, K., Andrawes, K.Z. (eds.) Proceedings of the International Reinforced Soil Conference Held in Glasgow, Scotland, Performance of Reinforced soil Structures vol. 43, pp. 471–473. Thomas Telford (1990)

Maher, M.H., Ho, Y.C.: Mechanical properties of kaolinite/fibre soil composite. J. Geotech. Geoenviron. Eng. **120**(8), 1381–1393 (1994)

Jayanthi, P.N.V., Singh, D.N.: Utilization of sustainable materials for soil stabilization. **5**(1) (2016). https://doi.org/10.1520/acem20150013

Kumar, P.: Stabilization of dune sand mixed with plastic (LDPE) waste strips for design of flexible pavement in construction of roads. Am. J. Eng. Res. (AJER) **5**(11), 315–320 (2016)

Bhattarai, P., Bharat Kumar, A.V.A., Santosh, K., Manikanta, T.C., Tejeswini, K.: Engineering behavior of soil reinforced with plastic strips. Int. J. Civ. Struct. Environ. Infrastruct. Eng. Res. Dev. (IJCSEIERD), **3**(2), 83–88 (2013). ISSN 2249-6866, TJPRC Pvt Ltd

Rathan Raj, R.: Stabilization of soil using rice husk ash. Int. J. Comput. Eng. Res. (IJCER) **6**(2) (2016)

Tiwari, S.K., Sharma, J.P.: Characterization and strength improvement techniques for collapsible dune sands. EJGE **18** (2013)

Rajput, S.S.: Effect of fly ash on geotechnical characteristics of Black cotton soil. IJIRST – Int. J. Innov. Res. Sci. Technol. **2**(3) (2015)

Comparison of Performance of Various Vertical Drains

Fifi Jacob[1], Anil Joseph[2(✉)], S. Chandrakaran[2], N. Sankar[2], and Babu T. Jose[3]

[1] Department of Civil Engineering, Albertian Institute of Science and Technology, Cochin University P.O, Kochi, Kerala, India
fifijacobp@gmail.com
[2] Department of Civil Engineering, National Institute of Technology, Calicut 673601, Kerala, India
aniljoseph01@gmail.com, {chandra, sankar}@nitc.ac.in
[3] Albertian Institute of Science and Technology, Cochin University P.O, Kochi, Kerala, India
babutjose@ymail.com

Abstract. India has a vast extend of coastal area and has got large deposits of soft marine clays extending to a depth of 15–60 m. Construction on marine clays is a big challenge because of its high compressibility, low shear strength and excessive settlement. Ground Improvement techniques such as sand drain, gravel drains, PVD, stone columns etc. are widely used to improve the engineering properties of sub soil. It was found from field studies that in case of very soft clay, if sand or coarse aggregate alone is provided, diameter of drain gets reduced and continuity of drain get broken, due to necking action of very soft clay. In order to reduce these effects and to increases strength and continuity characteristics of drainage media a hybrid mix comprising of aggregate, sand and synthetic fibre is developed through experimental model studies. The major limitation of displacement vertical drains is the remoulding of soil surrounding each drain caused by installation which reduces permeability and efficiency of drain. The problems of smear zone can be reduced by multiple driving of vertical drains in the same spot which will enhance efficiency of drains by forming new flow paths, as there is increase in diameter and drainage media get penetrated into smear zone. In this paper efficiency comparison of various vertical drains such as sand drain (single and double driving), PVD, and hybrid mix drain (single and double driving) is studied by assessing the behavior in laboratory on large scale model bed tanks. The efficiency is evaluated by comparing settlement with time for different loadings for each types of vertical drain. From the results it is noted that double driven hybrid mix shows much better performance compared with other vertical drains and can be effectively used for very soft clays.

Keywords: Vertical drains · Hybrid mix · Settlement · Smear zone

H. Shehata and H. Poulos (Eds.): GeoMEast 2018, SUCI, pp. 85–100, 2019.
https://doi.org/10.1007/978-3-030-01917-4_7

1 Introduction

Large scale consolidation of soft marine clays causes numerous problems in constructions and often affects the safety and performance of the structures. Vertical drains have been used extensively to accelerate the consolidation process of soft clay sub-soil by preloading. In the 1930s the first reasonable application of vertical drains was made in California. In Sweden, during the same decade, Kjellman introduced the first prototype. Subsequently, several types of vertical drains were developed, which consists of longitudinal channel wick functioning as a drain.

Barron (1948) presented the most comprehensive solution to the problem of radial consolidation by drain wells. The installation of vertical drains by means of a mandrel by displacement system of columnar intrusion causes significant remolding of the subsoil, especially in the immediate vicinity of the mandrel. Hansbo (1979) introduced a zone of smear in the vicinity of the drain with a reduced value of permeability. The presence of smear zone significantly influences the horizontal consolidation rate and is reported by various researchers (Onoue et al. 1991; Hird and Moseley 2000; Sharma and Xiao 2000; Joseph et al. 2015). Muttharam and Mohanapriya (2016) compared the rate of acceleration of consolidation by providing different drains such as sand drains and PVD. Rangeard et al. (2016) studied the influence of the sand column installation method on settlement reduction. Also they analyze the soil consolidation that takes place around the column during installation and its influence on the soil-column composite's behaviour. Redana (1999) conducted large scale radial drainage consolidation test on reconstituted Muar clay with and without sand compaction pile and prefabricated vertical drain. Deng et al. (2016) conducted laboratory model test on soft soil foundations with and without a prefabricated vertical drain (PVD) to assess the consolidation and long-term performance of PVDs.

Indraratna et al. (2009) conducted cyclic load test to examine the effectiveness of a prefabricated vertical drains (PVD) installed in reconstituted soft clay, large-scale cyclic triaxial tests was conducted on soft kaolinite clay with and without PVD. Tran-Nguyen et al. (2014) conducted laboratory performance test to study the discharge capacity behaviour of deformed PVDs. Samadhiya et al. (2008) conducted model load test on clay bed with and without granular piles and with and without nylon fibres. Al. Refeai (1992) conducted triaxial compression test on composite soil specimen made of annular soft clay samples, with both fiber reinforced and unreinforced sand columns. Andreou et al. (2008) conducted experimental study on the strength characteristics of foundation soil reinforced with compacted sand column and gravel column. Iai et al. (1988) conducted large scale model tests on gravel drains, installed as one of remedial measures against liquefaction. Joseph et al. (2016) presented the case study of ground improvement of embankments by multiple driven fibre reinforced columnar intrusions for vertical drains.

This paper presents the comparison of efficiency for M sand piles with single driving, M sand piles with double driving, optimum hybrid mix with single driving, optimum hybrid with double driving and PVD by assessing the behavior in the laboratory on large scale model bed tanks.

2 Materials

- Soft Clay

The soft clay used for the study was collected from a site near Vallarpadom container terminal, Cochin. The clay was collected from 8 to 9 m depth and kept in polythene bags to preserve the water content and transported to the laboratory. The physical properties of the soil are determined as per Indian standard specifications. The soil is classified as CH indicating that it is silty clay of high plasticity. The basic properties of marine clay presented in Table 1.

Table 1. Basic properties of marine clay

Sl no.	Properties		Value
1	Natural moisture content,%		112
2	Specific gravity		2.62
3	Grain size distribution	%sand	2
		%silt	54
		%clay	44
4	Liquid limit,%		156
5	Plastic limit,%		34
6	Plasticity index,%		122
7	Shrinkage limit,%		10.71
8	pH		7.53
9	Coefficient of consolidation (m^2/s)		5.65×10^{-8}
10	Void ratio		2.32
11	Compression index		0.64
12	Unconfined compressive strength, kPa		4

- River Sand

The river sand obtained from Periyar river was used for providing a thin central drain. A central drain of 10 mm in diameter, was formed with filter paper at center of all beds to allow radial drainage. Sieve analysis was conducted on the river sand and is classified as poorly graded sand (SP). The basic properties of river sand are determined and shown in Table 2.

- Manufactured Sand (M sand)

The manufactured sand used for the study was collected from Malayattor. Sieve analysis was conducted on the M sand and is classified as poorly graded sand (SP). The basic properties of M sand are determined and shown in Table 3.

Table 2. Basic properties of river sand

Sl. no.	Properties	Values	Test method
1	Specific gravity	2.576	IS 2720 Part 3 1980
2	Effective size, D_{10}	0.18	IS 2720 Part 4 1985
3	Uniformity coefficient	3.5	IS 2720 Part 4 1985
4	Coefficient of curvature	1.08	IS 2720 Part 4 1985
5	Max void ratio	0.961	IS 2720 Part 14 1983
6	Min void ratio	0.652	IS 2720 Part 14 1983
7	Average particle size, D_{50}	0.5	IS 2720 Part 4 1985
8	Void ratio, e	0.837	
9	% Fines	0.04	
10	Classification	SP	IS 1498 - 1970

Table 3. Basic properties of manufactured sand

Sl. no.	Properties	Values	Test method
1	Specific gravity	2.73	IS 2720 Part III 1980
2	Effective size, D_{10}	0.075	IS 2720 Part IV 1985
3	Uniformity coefficient	16	IS 2720 Part IV 1985
4	Coefficient of curvature	0.87	IS 2720 Part IV 1985
5	Max void ratio	0.72	IS 2720 Part XIV 1983
6	Min void ratio	0.40	IS 2720 Part XIV 1983
7	Average particle size, D_{50}	0.70	IS 2720 Part IV 1985
8	Void ratio, e	0.592	
9	% Fines	0.124	

- Coarse Aggregates

The coarse aggregates used for the study is of 12 mm downgraded size aggregate obtained from crushing of granitic rock. The basic physical properties of coarse aggregates are presented in Table 4.

Table 4. Basic properties of coarse aggregate

Sl. no.	Properties	Values	Test method
1	Specific gravity	2.78	IS 2386 (Part III) 1963
2	Water absorption %	0.69	IS 2386 (Part III) 1963
3	Bulk relative density (kN/m^3)	15.33	IS 2386 (Part III) 1963

- Synthetic Fibre

Synthetic fibre used for the study is Recron 3S fibre and is shown in Fig. 1. It is a secondary reinforcement product for construction. The uniqueness of synthetic fibre is its triangular shape, which give better anchorage in the mix. Also it is hydrophobic and chemically inert in nature. The properties of synthetic fibre are given in Table 5.

Fig. 1. Synthetic fibre – Recron 3S

Table 5. Basic properties of synthetic fibre

Particulars	Unit	Specified value
Appearance	–	White shiny glass roving
Tex	g/km	1000 (±5%)
LOI	%	Max 1.0
Diameter	mm	16
Moisture	% Max.	0.2
Tensile strength	N/tex	Min. 0.5

Note: Tex: Weight in grams of 1000 m of fiber or yarn.
N/tex: Unit of specific stress (force/linear density)
LOI: Limiting Oxygen Index

- Hybrid Mix as Drainage Media

It is found from the laboratory experiments that in case of very soft clay if sand or aggregate alone is used as drain media the diameter of the well get reduced due to the necking action and the clogging of the well is likely to be high. In order to reduce the effect of this phenomenon and to increase the strength and performance characteristics of the drainage media, a hybrid mix comprising of 12 mm down size aggregate, M sand and fibre reinforcements was developed. A series of direct shear tests were performed to obtain strength parameters, of various materials such as sand, aggregate and fibre in different proportions and to arrive at the optimum mix based on strength criteria. By installation of vertical drains in soft cohesive soil, the drainage path will reduces and the consolidation process of soft cohesive soil accelerates. The drainage depends upon the permeability of drain material. In order to find out the hydraulic characteristics of hybrid mix at various proportions, a series of permeability test were conducted. Also the evaluation of the capability of the hybrid content to self stand was determined using the slump cone test apparatus. The tests were conducted on various materials such as sand, aggregate, and fiber, with various proportions and the slump and horizontal flow of each hybrid mix were measured. The optimum hybrid mix obtained was 60% of 12 mm downgraded aggregates, 40% M sand and 0.03% synthetic fiber.

- Prefabricated Vertical Drain (PVD)

PVD of size 100 × 3.5 mm was used for this study. Equivalent diameter of drain was calculated using the equation 2 * (a + b)/π were "a" is the width and "b" is the thickness of drain and the equivalent diameter of drain obtained has 65.89 mm. Properties of PVD were collected from Manufacturer and are shown in Table 6.

Table 6. Properties of PVD

Sl. no.	Properties	Values	Test method
A	Overall drain		
1	Discharge capacity, m^3/s	$\geq 90 \times 10^{-6}$	ASTM D 4716
2	Tensile strength, N	≥ 2000	ASTM D 4595
3	Thickness, mm	>4	ASTM D 5199
4	Width, mm	100 ± 3 mm	ASTM D 5199
B	Filter		
1	Material	PP	
2	Colour	Black	
3	Permittivity test, s^{-1}	0.3	ASTM D 4491
4	Pore size/AOS, micron	120	ASTM D 4751
C	Core		
1	Material	PP	
2	Configuration	Corrugated	
3	Tensile strength, kN	1.3	ASTM D 4595
4	Compressive strength, kPa	1000	ASTM D 1621
5	Thickness, mm	3.5	ASTM D 5199

3 Efficiency Comparison of Various Vertical Drains

One of the major limitations of sand piling and PVD's is the smear zone generated due to the relative movement of the mandrel and the soil around. This leads to lower permeability within the smear zone. In addition, due to the surface tension formed on withdrawal of the mandrel, there will not be direct contact of the columnar material with the soft clay and unless a preload is applied drainage path, to dissipate the pore pressure formed in the clay will not be fully effective. It is noted that the forceful penetration of the hybrid material during the multiple driving at the same spot will decrease the loss in permeability in the smear zones, overcome the components of surface tension thereby commencing the dissipation of pore pressure without preloading and reduce the well resistance. Also inorder to reduce the clogging action of drainage media and necking action of well a hybrid mix comprising of 60% of 12 mm downgraded aggregates, 40% M sand and 0.03% synthetic fiber (Recron 3S) was developed based on direct shear test, permeability test and slump cone test studies. The efficiency comparison of vertical drains such as sand drain (single and double driving), PVD, and hybrid mix drain (single and double driving) is studied by assessing the behavior in laboratory on large scale model bed tanks. The efficiency is evaluated by comparing settlement with time for different loadings for each types of vertical drain.

3.1 Experimental Setup

Circular tanks of 600 mm diameter and 500 mm height were used as large scale model tanks. The tanks were made in such a way that it can be removed into 5 pieces, so that if required, the cross section could be inspected visually at any depth and to extract samples from various depths in horizontal as well as vertical directions. After the assembling of the tank by stacking the 5 pieces one over the other, the keys were tightened for locking. The friction along the tank walls was reduced by placing 1.5 mm thick polythene sheet around the internal tank boundary.

3.2 Preparation of Soil Sample

As it was not feasible to obtain an undisturbed sample for the large scale consolidation apparatus, reconstituted marine clay was used to make model bed tanks. Water was added to soft marine clay to take the water content, very close to liquid limit. It is kept covered with impermeable plastic sheet for one day to ensure uniform water content, after one day clay sample was mixed well and checked for liquid limit.

3.3 Preparation of Reconstituted Bed for Columnar Intrusion

After the assembling of the tank by stacking the 5 pieces one over the other with keys tightened for locking, it was filled with soft marine clay. Care was taken to ensure that no air bubbles are trapped in the clay mass. A thin central drain of sand, 10 mm in diameter, was formed with filter paper at center of all beds to allow radial drainage as shown in Fig. 2.

All the 6 model tanks were subjected to an initial pre-consolidation pressure of 2.5 kPa for 4 weeks in order to prepare a standard test bed of very soft clay layer. Once all the reconstituted beds were stabilized, 1st tank was kept as model bed and vertical drains were installed in the remaining 5 tanks.

Fig. 2. Preparation of reconstituted beds for installation of vertical drains

3.4 Installation of Various Vertical Drains

After stabilizing the reconstituted bed, columnar intrusion of M sand (single driving), was installed in tank no: 02, by using a pipe mandrel of 65 mm diameter and 2 mm thickness. Mandrel was driven vertically into the clay by a hammer of 26 N falling from a height of 300 mm ensuring the verticality throughout the intrusion. Knowing the volume of sand pile and density, sand was measured and it was poured into the mandrel and rodded at every 100 mm layer. The mandrel was then slowly withdrawn carefully maintaining the verticality while applying light vibration to the pipe as well as compacting inner sand using a rod. The vertical drain thus formed can be regarded as a single driving sand compaction pile (columnar intrusion).

A novel procedure for improving the efficiency of vertical drain by multiple driving was attempted. Multiple driving at the same point will enhance the performance of columnar intrusions, since the material of columnar intrusion will be pushed into the smear zone formed during the initial installation process (Joseph et al. 2016). One of the major limitations of columnar intrusion is the smear zone generated during the installation of displacement vertical drains by means of mandrel. It causes significant remoulding of the subsoil especially in the immediate vicinity of the mandrel. Thus a zone of smear will be developed with reduced permeability and increased compressibility. It is expected that some of this disadvantages could be overcome by the forceful penetration of the intrusion material into the smear zone of the soft clay and the enhancement of permeability will lead to accelerated consolidation settlements and earlier strength gain for the clay. In tank no: 03, double driving sand compaction pile has been installed by driving the mandrel at the same point thereby pushing the initially introduced sand into the smear zone.

It is found from the field studies that in case of very soft clay if sand drain or stone column is provided the diameter of the well get reduced due to the necking action and the clogging of the well is likely to be high. In order to reduce the effect of this phenomenon and to increase the strength and performance characteristics of the drainage media, a hybrid mix comprising of 12 mm down size aggregate, M sand and fibre reinforcements was used. Also it was assumed that, when the fibers are present, it helps to find a path for the dissipation of excess pore pressure when the drainage path is closed. In tank no: 04, single driving hybrid mix columnar intrusion has been installed. In tank no: 05 double driving hybrid mix columnar intrusion was driven at the same point. In tank no: 06 PVD drain was installed using a mandrel. The size of the PVD drain was of 100 mm width and 3.5 mm thickness. The equivalent diameter works out to be 65 mm.

After the installation of various vertical drains, in the five tanks, all the six tanks were subjected to consolidation and load of 5, 10, 15, and 20 kPa was applied for each tank in 4 stages. For each load increment settlement variations are measured with dial gauges and after the consolidation for a period of two weeks was completed the next increment of pressure is applied and the performance is evaluated. The process of installation of vertical drain and loading process is shown in Fig. 3.

Fig. 3. Installation of vertical drains in model beds and loading in progress

4 Test Results and Discussions

The comparison of efficiency for Model Bed, M sand piles with single driving, M sand piles with double driving, optimum hybrid mix with single driving, optimum hybrid mix with double driving, and PVD was determined by assessing their behaviour in the laboratory on large scale model bed tanks.

Figure 4 gives the variation in settlement with days for the different model beds for 5 kPa. From the graph it is noted that compared with the model bed with no vertical drain PVD drain shows 441% more settlement, M sand single driving shows 390% more settlement, hybrid single driving shows 553% more settlement, M sand double driving shows 581% more efficiency and hybrid double driving shows 699% more efficiency. The comparison of efficiency for M sand piles with single driving and M Sand piles with double driving shows a faster consolidation in case of double driving which is about 39% higher. PVD bed shows a 10% more settlement compared with single driven M sand and the double driven M sand bed shows a 26% higher settlement compared with PVD. PVD bed shows a 21% lesser settlement compared with single driven hybrid mix and the double driven hybrid mix shows a 48% higher settlement compared with PVD. Double driven hybrid mix shows a 17% higher settlement compared with double driven M sand.

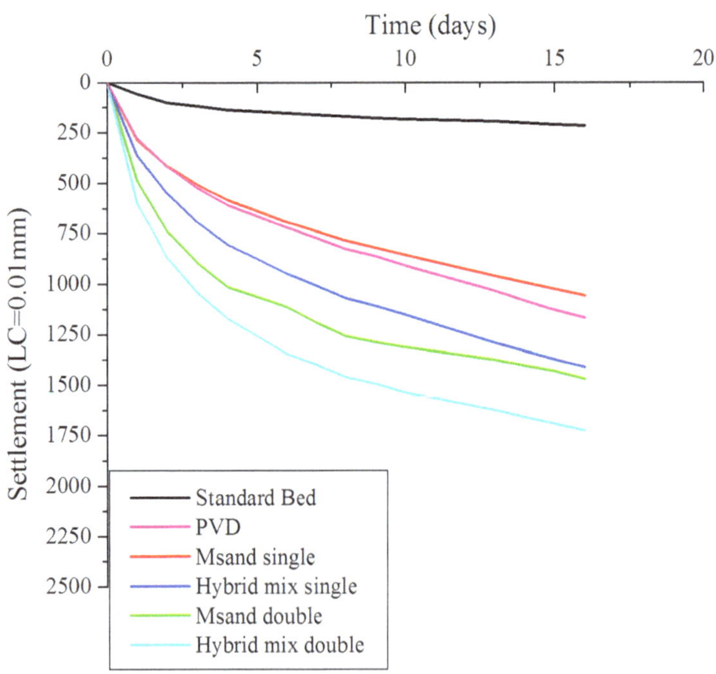

Fig. 4. Consolidation curve for different model beds under an application of 5 kPa pressure.

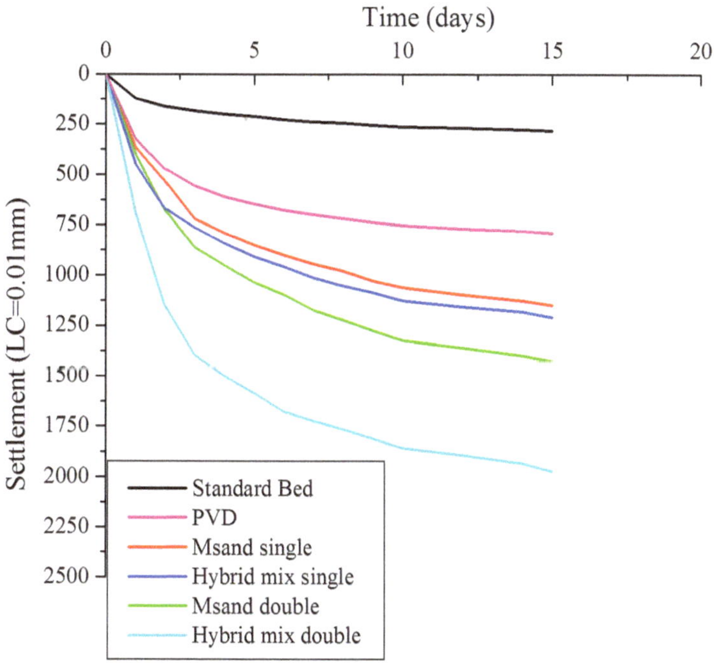

Fig. 5. Consolidation curve for different model beds under an application of 10 kPa pressure.

Figure 5 gives the variation in settlement with days for the different model beds for 10 kPa. From the graph it is noted that compared with the model bed with no vertical drain PVD shows 179% more settlement, M sand single driving shows 305% more settlement, hybrid single driving shows 326% more settlement, M sand double driving shows 403% more efficiency and hybrid double driving shows 595% more efficiency. The comparison of efficiency for M sand piles with single driving and M sand piles with double driving shows a faster consolidation in case of double driving which is about 24% higher. PVD bed shows a 45% less settlement compared with single driven M sand and the double driven M sand bed shows a 80% higher settlement compared with PVD. PVD bed shows a 53% lesser settlement compared with single driven hybrid mix and the double driven hybrid mix shows a 149% higher settlement compared with PVD. Double driven hybrid mix shows a 38% higher settlement compared with double driven M sand.

Figure 6 gives the variation in settlement with days for the different model beds for 15 kPa. From the graph it is noted that compared with the model bed with no vertical drain PVD shows 111% more settlement, M sand single driving shows 108% more settlement, hybrid single driving shows 159% more settlement, M sand double driving shows 203% more efficiency and hybrid double driving shows 233% more efficiency. The comparison of efficiency for M sand piles with single driving and M sand piles with double driving shows a faster consolidation in case of double driving which is about 45% higher. PVD bed shows almost same settlement compared with single

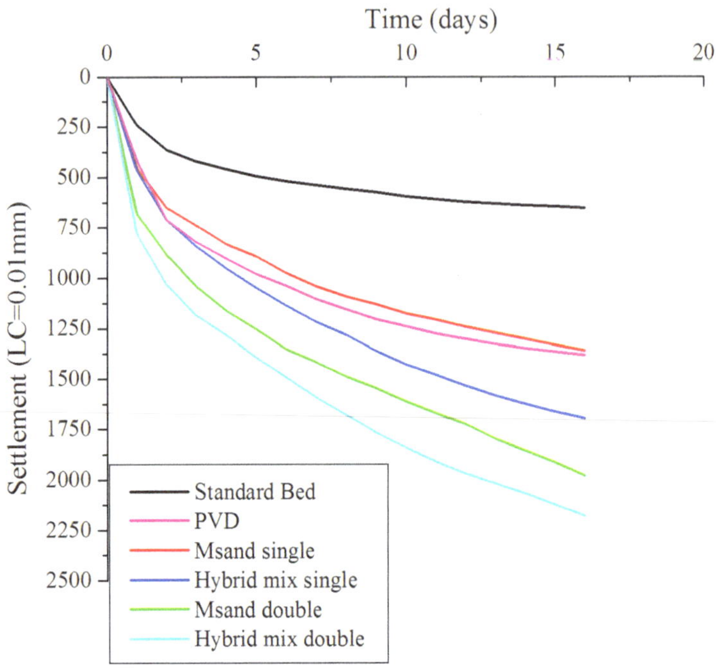

Fig. 6. Consolidation curve for different model beds under an application of 15 kPa pressure.

driven M sand and the double driven M sand bed shows a 43% higher settlement compared with PVD. PVD bed shows a 22% lesser settlement compared with single driven hybrid mix and the double driven hybrid mix shows a 57% higher settlement compared with PVD. Double driven hybrid mix shows a 10% higher settlement compared with double driven M sand.

Figure 7 gives the variation in settlement with days for the different model beds for 20 kPa. From the graph it is noted that compared with the model bed with no vertical drain PVD shows 60% more settlement, M sand single driving shows 82% more settlement, hybrid single driving shows 89% more settlement, M sand double driving shows 170% more efficiency and hybrid double driving shows 229% more efficiency. The comparison of efficiency for M sand piles with single driving and M Sand piles with double driving shows a faster consolidation in case of double driving which is about 48% higher. PVD bed shows 14% lesser settlement compared with single driven M sand and the double driven M sand bed shows a 69% higher settlement compared with PVD. PVD bed shows a 18% lesser settlement compared with single driven hybrid mix and the double driven hybrid mix shows a 106% higher settlement compared with PVD. Double driven hybrid mix shows a 22% higher settlement compared with double driven M sand.

Figure 8 shows the variation in settlement with days for various vertical drains and Fig. 9 shows variation of settlement with pressure for various vertical drains. From the

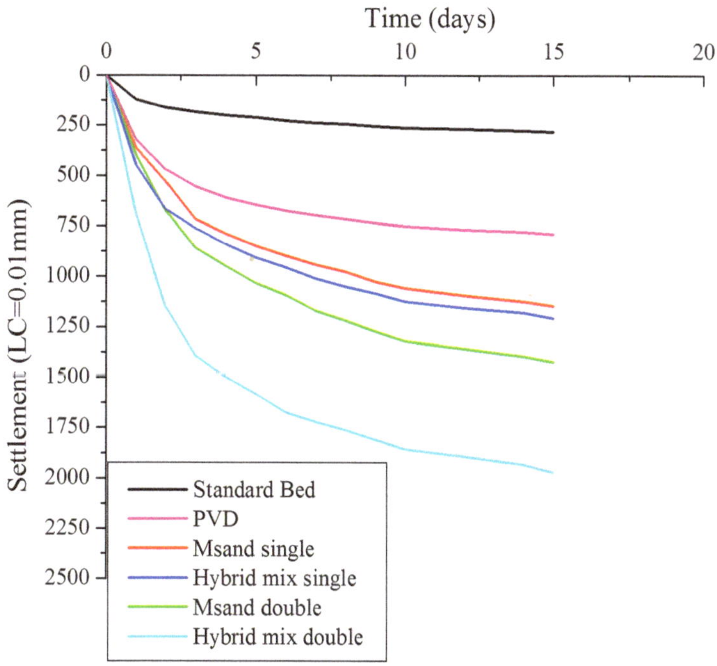

Fig. 7. Consolidation curve for different model beds under an application of 20 kPa pressure.

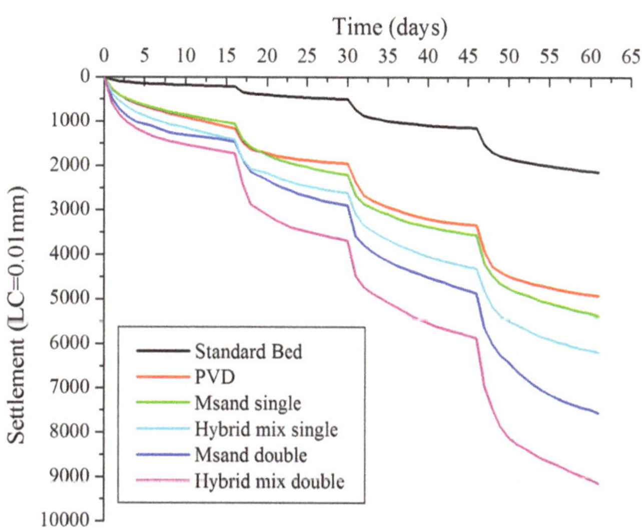

Fig. 8. Variation of settlement with days for various vertical drains

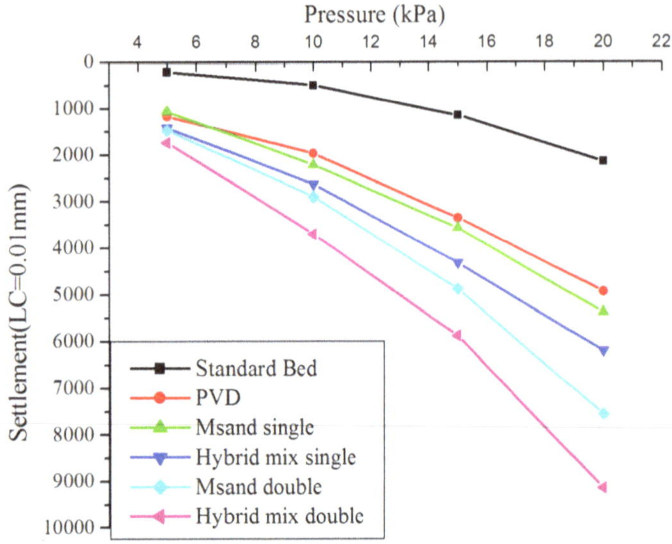

Fig. 9. Variation of settlement with pressure for various vertical drains

result it is observed that the maximum consolidation settlement is achieved in case of double driven hybrid mix. Compared with the model bed with no vertical drain on the cumulative application up to 20 kPa pressure PVD shows 130% more settlement, M sand single driving shows 151% more settlement, hybrid single driving shows 188% more settlement, M sand double driving shows 252% more efficiency and hybrid double driving shows 327% more efficiency.

5 Conclusions

1. The comparison of efficiency for M sand piles with single driving and M sand piles with double driving shows a faster consolidation in case of double driving which is about 41% higher.
2. PVD bed shows 9% lesser settlement compared with single driven M sand and the double driven M sand bed shows a 53% higher settlement compared with PVD.
3. PVD bed shows a 26% lesser settlement compared with single driven hybrid mix and the double driven hybrid mix shows a 85% higher settlement compared with PVD. Double driven hybrid mix shows a 21% higher settlement compared with double driven M sand.
4. The improved performance of the multiple driven beds may be due to the forceful penetration of the intrusion material into the smear zone of the soft clay and the enhancement of permeability which has led to accelerated consolidation settlements and earlier strength gain for the clay.
5. It is presumed that the forceful penetration of the fibre reinforced columnar intrusion during the second driving at the same spot will help overcome the components

of surface tension there by commencing the dissipation of pore pressure without preloading and reduce the well resistance.

6. The hybrid mix enhanced performance might be due to the optimum aggregate, M sand proportion which improves the permeability and the presence of synthetic fibers, which will act as a reinforcement, prevents cracks, necking failure and ensure continuity of the drains.

References

Al-Refeai, T.O.: Strengthening of soft soil by fiber reinforced sand column. In: Proceedings of International Symposium on Earth Reinforcement Practice, Fukoka, Japan, vol. 1, pp. 677–682 (1992)

Andreou, P., et al.: Experimental study on sand and gravel columns in clay. J. Ground Improv. Proc. Inst. Civ. Eng. **161**(4), 189–198 (2008)

ASTM (American Society for Testing and Materials): Standard test method for determining the (in-plane) flow rate per unit width and hydraulic transmissivity of a geosynthetic using a constant head, Pennsylvania, United States

ASTM (American Society for Testing and Materials): Standard test method for tensile properties of geotextiles by the wide-width strip method, Pennsylvania, United States

ASTM (American Society for Testing and Materials): Standard test method for measuring the nominal thickness of geosynthetics, Pennsylvania, United States

ASTM (American Society for Testing and Materials): Standard test methods for water permeability of geotextiles by permittivity, Pennsylvania, United States

ASTM (American Society for Testing and Materials): Standard test method for compressive properties of rigid cellular plastics, Pennsylvania, United States

ASTM (American Society for Testing and Materials): Standard test methods for determining apparent opening size of a geotextile, Pennsylvania, United States

Barron, R.A.: Consolidation of fine-grained soils by drain wells. Trans. Am. Soc. Civ. Eng. **113** (1), 718–742 (1948)

Deng, Y.B., et al.: Model test and theoretical analysis for soft soil foundations improved by prefabricated vertical drains. Int. J. Geomech. ASCE **17**(1), 04016045 (2016)

Hansbo, S.: Consolidation of clay by band shaped prefabricated drains. Ground Eng. **12**(5), 16–25 (1979)

Hird, C.C., Moseley, V.J.: Model study of seepage in smear zones around vertical drains in layered soil. Geotechnique **50**(1), 89–97 (2000)

Iai, S., Koizumi, K., Noda, S., Tsuchida, H.: Large scale model tests and analysis of gravel drains. In: Proceedings of 9th World Conference on Earthquake Engineering, vol. 3, pp. 261–266 (1988)

Indraratna, B., et al.: Experimental investigation on effectiveness of a vertical drain under cyclic loads. J. Geotech. Geo environ. Eng. ASCE **135**(6), 835–839 (2009)

Joseph, A., Chandrakaran, S., Sankar, N., Jose, B.T.: Laboratory evaluation of extent of smear zone due to columnar intrusion for cochin marine clays. In: 50th Indian Geotechnical Conference, Pune (2015)

Joseph, A., Jose, B.T., Chandrakaran, S., Sankar, N.: Multiple driven fibre reinforced columnar intrusions for vertical drains - a case study. In: Indian Geotechnical Conference, Chennai (2016)

Muttharam, M., Mohanapriya, A.: Study on rate of consolidation of soft clay using sand drain and PVD. In: Indian Geotechnical Conference, Chennai (2016)
Onoue, A., Ting, N.H., Germaine, M., Whitman, R.V.: Permeability of disturbed zone around vertical drains. In: Proceedings of ASCE Geotechnical Engineering Congress, vol. 2, pp. 879–890 (1991)
Rangeard, D., Phan, P.T., Martinez, J., Lambert, S.: Mechanical behavior of fine-grained soil reinforced by sand columns: an experimental laboratory study. Geotech. Test. J. **39**(4), 648–657 (2016)
Redana, W.I.: Effectiveness of vertical drains in soft clay with special reference to smear effect. Research online, University of Wollongong, Australia, Ph.D. thesis (1999)
Samadhiya, N.K., et al.: Load settlement characteristics of granular piles with randomly mixed fibres. Indian Geotech. J. **38**(3), 345–354 (2008)
Sharma, J.S., Xiao, D.: Characterization of a smear zone around vertical drains by large-scale laboratory tests. Can. Geotech. J. **37**(6), 1265–1271 (2000)
Tran-Nguyen, H.H., et al.: Effect of deformation of prefabricated vertical drains on discharge capacity. Geo synth. Int. **17**(6), 431–442 (2014)

Stabilization of Subgrade Soil Using Polyacrylamide-Based Additive

Romel N. Georgees[1,2(✉)] and Rayya A. Hassan[1]

[1] Swinburne University of Technology, Melbourne, Australia
rgeorgees@swin.edu.au, eldadoner@yahoo.com
[2] Kirkuk University, Kirkuk, Iraq

Abstract. In Australia, polymers have been used as pavement material stabilizer or dust suppressant for unsealed roads. However, the use of polymers, as stabilizers of pavement materials, has not been fully accepted by Australian Road Authorities as a standard practice since they do not have relevant design processes. Therefore, this study has been carried out to better understand the benefits of using a polyacrylamide-based additive (PAM) in stabilizing pavement construction materials. In this study, a synthetic PAM was used to stabilize three types of soil used in unsealed pavement construction. Unconfined Compressive Strength, California Bearing Ratio and erosion tests were conducted to evaluate the performance of soils treated with PAM. The results showed a significant overall increase in strength for all soil types with varying degrees. The results also showed improvements in bearing capacity of all the treated soils and their erosion resistance.

Keywords: Pavement materials · Subbase · Unconfined compressive strength California Bearing Ratio · Erosion resistance

1 Introduction

Australian State and Federal governments are under pressure to design, build and maintain road infrastructure within the conservative funds available. Furthermore, the public are constantly demanding higher performing roads that are environmentally sustainable (via using more recycled materials or improved marginal quality materials). This has created a focus on the development of innovative construction techniques that allow low quality local materials to be used in road pavement construction. This is particularly important in the urban environment where high quality aggregates are scarce and associated transport costs are high.

The stabilization of pavement materials using cementitious additives has proven to be a successful and reliable technique to increase the performance of pavement materials and prolong the service life of a pavement structure. However, to be effective, large quantities of such additives are required as well as relatively long curing times. This will significantly increase the construction time and associated costs. The other drawback of using a cementitious type additive is that it increases the risk of reflective cracking in the asphaltic surface layer (Judy et al. 2005; Rodway 2001). Therefore, the development of non-traditional stabilizers, such as polymers, have gained significant

H. Shehata and H. Poulos (Eds.): GeoMEast 2018, SUCI, pp. 101–111, 2019.
https://doi.org/10.1007/978-3-030-01917-4_8

attention recently as they have demonstrated effectiveness in the field with regards to reducing permeability, increased durability, and being non-time dependent during mixing and compaction, as well as being more environmentally sustainable (Camarena 2013; Wilmot 1994).

The principal function of most of these polymers is to provide a distinctive hydrophobic characteristic to the treated layer to preserve an adequate dry strength for moisture-susceptible materials, in addition to providing a bonding action between particles (Wilmot and Rodway 2000). Polyacrylamide (PAM), which is a hydrophilic polymer, has been used in the field of agriculture for nearly half a century due to its ability to increase infiltration and decrease irrigation-induced erosion (Sojka et al. 2007). Yet, until recently its use to stabilize pavement materials has been very rare.

Deng et al. (2012) tested the effectiveness of a polyacrylamide-based additive on changing the engineering behaviour of a highly plastic clay soil with the aim of establishing the failure mode of the soil treated with PAM. It was concluded that the addition of PAM showed an increase in failure strain of the plastic clay soil, which increased its flexibility. Another extensive experimental test by Roa-Espinosa and Mikel (2004) was carried out to assess the effect of PAM in reducing dust generation in a sandy soil of helicopter lands at the Marine Corps Air Ground Combat Center (MCAGCC) in California, USA. Roa-Espinosa and Mikel (2004) concluded that PAM was effective in reducing dust generation, when comparing treated plots with the untreated ones.

The only available published Australian study to evaluate the effect of PAM on the durability of pavement material was performed in the field (Andrews and Sharp 2010). The study comprised of evaluating an unsealed road pavement after being stabilized with PAM. The treated pavement section was 400 m long and 8 m wide and comprised of silty sand soil. Performance of the PAM-stabilized pavement was monitored over 18 months. It was found that the untreated pavement sections deteriorated rapidly over the first 12 months due to environmental and traffic conditions, whereas the treated sections maintained their structural integrity throughout the entire 18 months monitoring period. In particular, the surface of the untreated section developed into a rough and open texture with many loose particles, while the treated section showed a tighter texture with very few loose particles.

Due to the success of the abovementioned field trial, an extensive and systematic laboratory investigation was performed to assess the potential durability improvements of PAM-treated materials used for select fills or granular subbase materials. The additive used in this study was an "off-the-shelf" product that is currently available in Australia and has demonstrated positive results in the field.

2 Materials and Testing Methods

Three different types of soils from three different sites in Australia were selected and referred to herein as soils A, B and C. These soils were retrieved form in-situ stabilization projects. The samples typically comprised of up to 20 mm aggregate size.

The polymeric additive used in this study was a synthetic soluble anionic polyacrylamide (PAM), which is produced in granular form. The PAM has a moderate charge density of about 18% and a high molecular weight of typically 12–15 Mg/mol. The product is a non-toxic water-soluble material with a specific gravity of 0.8 and a PH value of 6.9 at 25 °C.

Physical characterization tests of the soils, including plasticity limit, liquid limit and plasticity index as well as particle size distribution were performed using Australian Standards. Classification of the soils according to unified soil classification system (USCS) were then determined (ASTM 2011). The geotechnical properties as well as classifications of these soils according to the USCS, along with particle size proportions and the above indices are tabulated in Table 1. The particle size distribution plots of these soils are presented in Fig. 1.

Table 1. Properties of soil types A, B and C

Properties	Soil A	Soil B	Soil C
% Gravel (4.75–19.0 mm)	33.2	29.6	13.2
% Sand (0.075–4.75 mm)	56.8	46.4	36.8
% Fines (<0.075 mm)	10.0	24.0	50.0
% Clay (<2 μm)	0.85	2.75	13.5
Liquid limit (%)	22.2	23.8	31.4
Plastic limit (%)	N/A	12.8	15.1
Plasticity index (%)	N/A	11	16.3
Optimum moisture content (%)	5.6	8.5	12.5
Max dry density (g/m^3)	2.35	2.01	1.96
Water absorption: average (%)	3.2	16.5	6.8
Soil classification, (USCS)	(SP-SM)	(SC)	(CL)

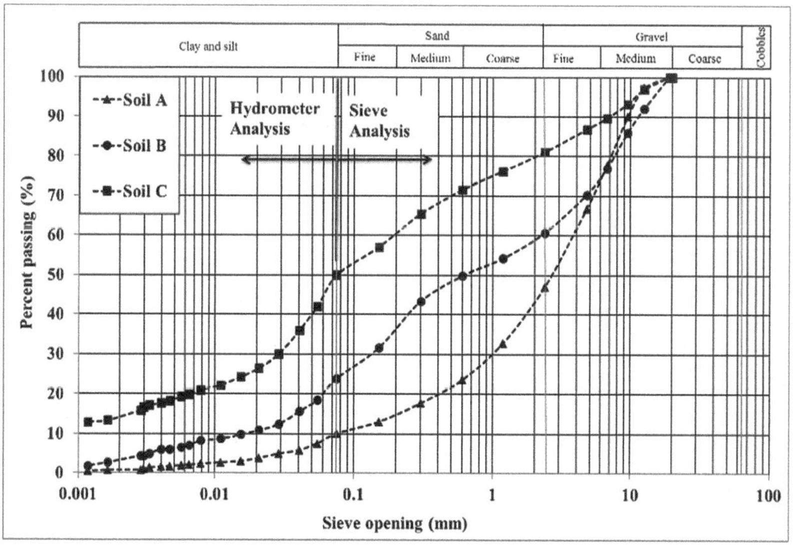

Fig. 1. Particle size distribution of soils type A, B and C

Sieve analysis and hydrometer test results showed that the soils had a wide range of gradations ranging from well to medium and poorly graded for soil types A, B and C, respectively. According to the USCS classification system, soil type A is a coarse-grained soil of sand and gravel with silt (poorly graded, non-plastic), soil type B is a coarse-grained Clayey Sand with gravel (moderately plastic), and soil type C is a fine-grained Sandy clay with gravel. Referring to Table 1, one can conclude that the reason for the abnormal water absorption of soil type B (16.5%) is because of the highly porosity nature of soil.

Subsequently after physical characterization tests, unconfined compressive strength (UCS), California Bearing Ratio (CBR) and erosion resistance tests were undertaken in order to determine the shear strength property and sealing capacity of PAM in terms of reducing water erosion actions.

2.1 Sample Preparation

Laboratory test samples were prepared in accordance with relevant Australian standards. The required amount of PAM was determined according to the supplier's recommendation, which was 0.002% by dry weight of the soil. After mixing the soil-water mixture, samples were kept in sealed plastic bags and stored at room temperature for 24 h to allow even moisture distribution. The maximum dry density and optimum moisture content of treated and untreated specimens were determined using the Australian modified compaction test (AS 2003).

2.2 UCS Testing

For the UCS test, samples were compacted in Australian modified compaction moulds in five layers. However, based on preliminary experimental results for the treated soils with PAM, the optimum compaction efforts for the three soil types, were found to be 45 blows per layer (*i.e.* 4,961 kNm/m^3) for soil type A, and 35 blows per layer (*i.e.* 3,868 kNm/m^3) for soil types B and C. These compaction energies were then applied to both treated and untreated samples. These compaction efforts represent the optimum for each soil type. They were determined after a number of trials, including 25, 35 and 45 blows per layer, to ensure that the performance of laboratory prepared samples is comparable with relevant observed field performance.

Specimens were removed from the split moulds after compaction and stored in a curing room at a temperature of 25 ± 3 °C and relative humidity of 50 ± 0.5% until a constant mass was reached. The specimens were then sprayed with water containing dissolved PAM and left to dry back until a constant mass was reached. This procedure was used to simulate the sealing capacity of the polymer when used in the field, in which PAM is distributed evenly across the treated materials, but highly concentrated on the surface of the lift. These dried specimens were then subjected to a wet condition before being tested following the procedure outlined by Santoni et al. (2002). The only difference was the duration of soaking, in which a specimen was placed on its side in a

25.4 mm of water for 30 min rather than 15 min. After soaking, the specimen was left to drain for 5 min (to remove the excess water) before being tested via the UCS test. The specimens were then loaded either to the point where the load reached a maximum value and then decreased with increasing strain, or until 15% strain was reached (ASTM 2008). Typical sample (wet condition) is shown in Fig. 2.

Fig. 2. Wet sample (treated or untreated) during the UCS test.

2.3 CBR Testing

In this study the CBR testing was carried out to evaluate the potential improvement in bearing capacity of the three types of soils after being stabilised with PAM. Three specimens were prepared per sample (treated and untreated) in order to ensure reliable results. Anomalous results were discarded and replaced by other prepared samples. The test was conducted in accordance with Australian Standards AS 1289. 6.1.1 (AS 1998). The soils were compacted in a mould with a diameter of 152 mm and a height of 117 mm. The specimens were prepared by compacting five layers with the same optimal compactive energy as per the UCS specimens. The specimens were prepared according to dry condition. This was carried out when the specimens were left to dry back in air dry condition for 14 days. A surcharge of 4.5 kg loading was applied during the curing period to simulate pavement material loading above the tested layer (Vicroads 1998). Since the CBR test was conducted on dry condition, treated samples were not being sprayed with the dissolved PAM.

2.4 Erosion Testing

For the erosion resistance test, a method proposed by Jones and Ventura (2003) was used. Samples were compacted in Australian modified compaction moulds following the same optimum compaction efforts used for the preparation of UCS's specimens for the three soil types. The treated specimens were also sprayed with water containing dissolved PAM (after drying back to a constant mass) and left to dry back in an oven with 50 °C until a constant mass was reached. After drying back, specimens were then positioned in the erosion apparatus and subjected to a constant head of one meter of water flow to simulate the water flow on the road surface (Jones and Ventura 2003). The specimens were subjected to a water flow for 5 min and then allowed to drain for an extra 5 min. Three specimens were prepared per sample (treated and untreated). Specimens were then dried in an oven with 50 °C to a constant mass and the mass loss as a percentage to the original mass was recorded.

3 Results and Discussion

The effect of using PAM on the unconfined compressive strength for the three soil types was assessed using the wet condition test. The UCS values of the treated and untreated samples for the three soil types are presented in Table 2. It is worth noting that the strength values (treated and untreated samples) presented in the table represent an average UCS value based on at least three specimens.

Table 2. Wet UCS test results for the treated and untreated samples for soil types A, B and C.

Soil type	Sample size	Average UCS (MPa)	
		Untreated	Treated
A	3	1.4	2.6
B	3	1.26	3.11
C	3	1.42	2.76

Table 2 illustrates moisture susceptibility of the samples when treated with PAM and tested under wet conditions. Here, the retained strength (i.e. wet strength when tested under wet conditions) of the treated samples is dramatically higher when compared with the untreated samples. Soil types A and C show an increase in the retained strength of 85.6 and 94.3% respectively compared to the untreated ones. Soil type B, however, exhibits a 146.6% increase in strength when treated with PAM. This implies that treating soil with PAM provides a significant resistance to moisture deterioration, particularly for soils with higher porosity and water absorption (soil type B). It is believed that the polymer better sealed the soil's pores in a manner that prevented water permeating into the inner pores. The mechanism of preventing water entering the pores is related to increased viscosity of the water during contact between the polymer and soil surface. Thus, impeding water ingress and preserving the dry strength of the pavement material (Malik and Letey 1991).

The results of CBR tests of treated and untreated samples for the three soil types are presented in Table 3. The values presented in this table are the average CBR values of the three specimens per sample and their standard errors at 95% confidence level are also presented. It is worth noting that a CBR value greater than 100 is meaningless for pavement design, but such values are used herein for comparison purposes.

Table 3. CBR values for treated and untreated samples of soil types A, B and C

Soil type	Parameters		CBR, %		Percent increase
			Untreated	Treated	
A	35 BPL	Av.	188.3	230.5	22.4
		St. error	3.23	4.46	
B	35 BPL	Av.	467.6	499.1	6.8
		St. error	3.22	2.85	
C	35 BPL	Av.	13.8	18.9	37.3
		St. error	1.10	1.50	

In Table 3, a regular increase in the CBR values can be observed for all three soils treated with PAM when compared with the untreated ones. It also shows that the level of improvement varies with soil type. For instance, moderate increases in penetration resistance were observed in both treated soil types A and B at 3,868 kNm/m^3 compactive effort, while a remarkable increase in strength was recorded in the treated samples of soil type C. Changes in the bearing capacity of soil type A show an increase of 22.4% in favour of the treated samples. Whereas, treated samples of soil type B exhibited only 6.8% increase in CBR value compared with their untreated counterparts. The rate of change in CBR for soil B is considered relatively high when based on the high CBR values of the samples. On the other hand, a significant increase in CBR value was noted in the treated samples of soil C, with a maximum difference in CBR value of 37.3%. In fact, cohesion and frictional forces between contacted particles are the major parameters in determining the shear strength of the soil matrix. Therefore, PAM was considered as a major factor that enhanced the interlocking or adhering characteristics of soil particles, which certainly increased soil shear strength under CBR test.

A higher compactive effort was also applied to additional soil A samples to investigate the effect of increasing compactive effort on shear resistance property. The compactive energy was increased from 3,868 kNm/m^3 to 4,961 kNm/m^3. The results of change in the shear resistance is shown in Fig. 3. In fact, the test shows increased bearing capacity of the treated samples of soil type A by 17.2% when the compactive effort increased to 4,961 kNm/m^3. However, untreated samples show a slight decrease in CBR by approximately 3.6%. The difference between treated and untreated samples is 45.7%.

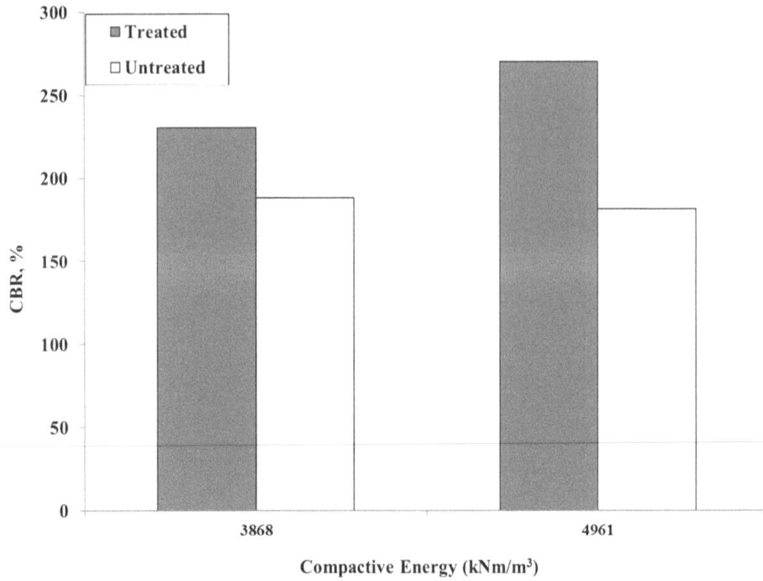

Fig. 3. Change in CBR values when changing compactive energy for treated and untreated samples of soil type A

The capacity of PAM to protect the compacted specimens from water flow is also tested using an erosion test. Figure 4 shows the potential effect of PAM additive to reduce the erosion action. In this figure, three specimens per sample (treated and untreated) were tested and the average values of mass loss (grams) are presented.

Figure 4 shows that the differences in mass loss, due to the erosion action, between the untreated and PAM treated samples for the three types of soils are evident. The figure also shows that there is a decrease in mass loss as the fines content increases. The greatest mass loss is recorded for soil type A and the least for soil type C. This is supported by the amount of plastic fines present in soil types A, B and C by 1, 2.75 and 13.5% respectively. As expected, the polymer increased the erosion resistance of the materials tested, which in turn increases the stability of the pavement layer after prolonged exposure to the rainfall during construction stages. A significant decrease in mass loss was found in the treated samples of soil type C at 94.5% followed by soil types A and B with 79.3 and 70.9%, respectively.

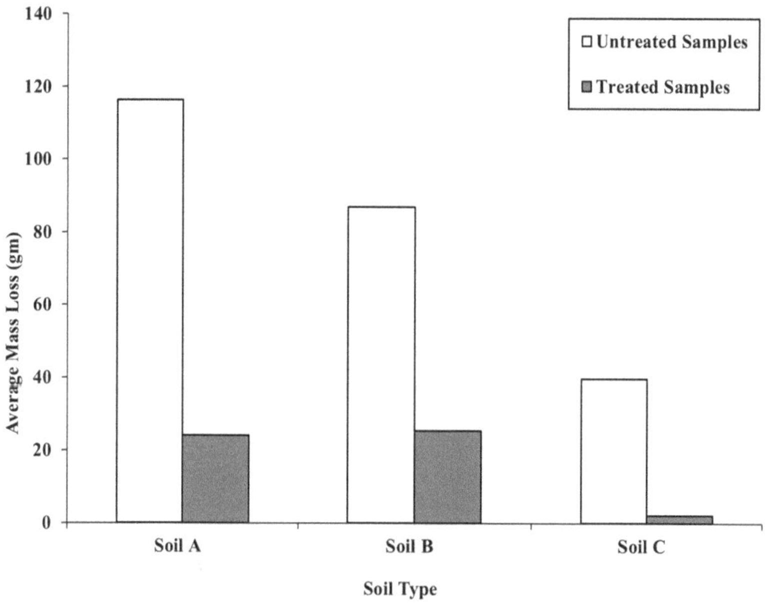

Fig. 4. Erosion resistance test for the treated and untreated samples of soil types A, B and C

4 Conclusion

From laboratory testing, the following has been concluded when using PAM as a stabilizing additive to treat unbound pavement materials:

- An increase in the wet strength was noted for all soils treated with PAM. The addition of PAM was found to improve soils' resistance to moisture absorption. The average increase in the retained strength of soil types A and C was 85.6 and 94.3% respectively compared to the untreated ones. Soil type B, on the other hand exhibited a 146.6% increase in strength when treated with PAM.
- Using PAM increased the unsoaked CBR for all the treated soils. The level of improvement was found to vary by soil type. The levels of improvements ranged from 6.8 to 37.3%. Soils with higher clay content revealed the greatest strength gain and soils with less fines content exhibited the least, when treated with PAM.
- An increase in the resistance to erosion was measured between 70.9 to 94.5% for all soil types treated with PAM. Here, the addition of PAM is more effective for the soils with greater fines and clay content. This was a significant result.

Overall, these results indicate that PAM is a promising stabilizing additive for sustainably improving the performance of pavement foundation materials with the potential of reducing pavement thickness which is important when rehabilitating urban pavements to match existing levels of other civil infrastructure such as the kerb and channel of drainage system. The increased strength will provide higher toughness structural layer, which mitigate the stresses imposed on the foundation by construction traffic during construction. Further, the improved durability and resistance to erosion of treated material helps to reduce loss of mass due to the erosion action of rainfall during the construction of the foundation unbound layers, in addition to facilitate staged construction to reduce traffic delay in urban environment.

Acknowledgement. The authors would like to acknowledge the contribution of Earthco projects Pty Ltd, particularly Mr. Mark Holding, for providing the stabilizer additive for this research.

References

Andrews, R., Sharp, K.: A protocol for conducting field trials for best value management of unsealed roads. In: 24th ARRB Conference – Building on 50 Years of Road and Transport Research, Melbourne, Australia, p. 14 (2010)

AS: Methods of testing soils for engineering purposes Method 6.1.1: soil strength and consolidation tests-determination of the California bearing ratio of a soil-standard laboratory method for a remoulded specimen, Sydney, Australia (1998)

AS: Methods of testing soils for engineering purposes Method 5.2.1: soil compaction and density tests—Determination of the dry density/moisture content relation of a soil using modified compactive effort, Sydney, Australia (2003)

ASTM: Standard Test Method for Unconfined Compressive Strength of Cohesive Soil, West Conshohocken, PA, USA (2008)

ASTM: Standard practice for classification of soils for engineering purposes (unified soil classification system), West Conshohocken, PA, USA (2011)

Camarena, S.: Sustainable road maintenance and construction utilising new technologies. In: International Public Works Conference, 2013, Darwin, Northern Territory, Australia (2013)

Deng, Y.F., Liu, S.Y., Huang, J., Kan, L., Du, Y.J., Jing, F.: Strength and permeability of cemented soil with PAM. In: Grouting and Deep Mixing 2012, pp. 1800–1807. ASCE (2012)

Jones, D., Ventura, D.: Development of fit-for-purpose certification testing for non-traditional road additives-Phases I-III. Contract Report: CR–2003/34, Pretoria: Division of Roads and Transport Technology, CSIR (2003)

Judy, F., Diliman, Q.C., Albert, C., O'Connell, M.J., Brs, E., John, R., House, C., Ride, N.M.: Stabilized sub-bases for heavily trafficked roads in the Philippines. J. East. Asia Soc. Transp. Stud. **6**, 1274–1285 (2005)

Malik, M., Letey, J.: Adsorption of polyacrylamide and polysaccharide polymers on soil materials. Soil Sci. Soc. Am. J. **55**(2), 380–383 (1991)

Roa-Espinosa, A., Mikel, A.: PAM application for dust mitigation in helicopter operations. Internal Rep. Marine Corps War Fighting Laboratory, Quantico, VA (2004)

Rodway, B.: Polymer stabilisation of clayey gravels. In: Proceedings for 20th ARRB Conference, Melbourne, Australia (2001)

Santoni, R.L., Tingle, J.S., Webster, S.L.: Stabilization of silty sand with nontraditional additives. Transp. Res. Rec. J. Transp. Res. Board **1787**, 61–70 (2002)

Sojka, R., Bjorneberg, D., Entry, J., Lentz, R., Orts, W.: Polyacrylamide in agriculture and environmental land management. Adv. Agron. **92**, 75–162 (2007)

Vicroads: Guide to general requirements for unbound pavement materials. Kew, VIC, Australia (1998)

Wilmot, T., Rodway, B.: Stabilised pavements-selecting the additive: cementitious, polymer or bitumen. Aust. Civ. Eng. Trans. **42**, 27–33 (2000)

Wilmot, T.D.: Applications for stabilisation in pavement construction and rehabilitation. In: Proceeding of the First Malaysian Road Conference: Roads-Towards Nation Building, p. 10 (1994)

Effect of Wet-Dry Cycles on Mechanical Strength Properties of Cement Stabilized Granular Lateritic Soil

Brundaban Beriha(✉), Dipti Ranjan Biswal, and Umesh Chandra Sahoo

School of Infrastructure, IIT Bhubaneswar, Bhubaneswar, India
{bb10, drb11, ucsahoo}@iitbbs.ac.in

Abstract. Scarcity of good quality stone aggregates has been posing a major challenge in sustainable development of roads in many developing countries including India. Granular lateritic soil is marginal material abundantly available in major parts of India and this needs to be stabilized for use in base and subbase layers. Traditionally durability is assessed through wet-dry durability test, where the specimens are subjected to alternate wetting and drying cycles, which is believed to simulate the natural weathering conditions. This work presented in this paper reports the effect of wet-dry (W-D) cycles on unconfined compressive strength (UCS), flexural strength (FS) and flexural modulus (FM) of a cement stabilized granular lateritic soil (CLS) collected from the eastern part of India. The strength and stiffness of CLS was found to be increasing with the number of durability cycles. This trend of increasing strength and stiffness is contradicting the general assumption that the W-D process reduces the material integrity and this may be attributed to the accelerated curing due to heating of samples, which negated the detrimental effect of wetting and also due to gradual strength gain with blended cements.

Keywords: Granular lateritic soil · Stabilization · Wet-dry durability test Unconfined compressive strength · Flexural strength · Flexural modulus

1 Introduction

The scarcity of natural aggregates has been posing a major concern for the construction of roads in many developing countries as they have to go a long way in establishing better road infrastructure. A possible relief for this concern is the use of marginal materials or locally available granular soils in structural layers of the pavements, which can reduce the demand for the use of natural aggregates. In India, granular lateritic soils are widely available in many parts of the country and presently, these are only used as subbase material in light trafficked pavements. These soils, when stabilized have the potential to replace the conventional granular base and subbase layers of highway pavements to address the material scarcity.

Any stabilized material needs to satisfy the minimum strength and durability criteria specified by different agencies across different countries e.g. National Lime Association (NLA 2006) for soil-lime, Portland Cement Association (PCA 1992) for

H. Shehata and H. Poulos (Eds.): GeoMEast 2018, SUCI, pp. 112–121, 2019.
https://doi.org/10.1007/978-3-030-01917-4_9

soil-cement, and the Federal Highway Administration (FHWA) (Veisi et al. 2010) for soil-fly ash. The strength criteria is satisfied by adding appropriate binder content to achieve a minimum specified unconfined compressive strength (UCS) value. The durability is assessed through wet-dry (W-D) durability test (ASTM D559 2015), where the specimens are subjected to alternate wetting and drying cycles, which is believed to simulate the natural weathering conditions. After specified durability cycles, the total mass loss is determined, which indicates the level of retention of integrity of the stabilized samples subjected to field weathering action. Many researchers (Ventura 2003; Scullion et al. 2005; Khoury and Zaman 2007) have raised concern over the credibility of this test method as it is too severe and depends on the judgment of the person performing the test. After investigating few trial sections, Kalankamary and Davidson (1963) and Bandara et al. (2015) have also reported that the brushing during durability test is too severe and does not simulate filed condition. Therefore, rather than measuring mass loss, changes in strength and stiffness parameters after each cycle should be assessed, which is more rational and can be used to study the mechanistic analysis of pavements. Several researchers (Khoury and Zaman 2002; Khoury and Zaman 2007; Zhang and Tao 2006) have studied the effect of W-D cycles on resilient modulus, unconfined compressive strength and modulus of elasticity of different stabilized material. But very few studies have been done to investigate the effect of durability cycles on cement stabilized granular lateritic soil (CLS). Hence, an experimental investigation was taken up in this work to study the effect of wet-dry cycles on different mechanical properties such as unconfined compressive strength (UCS), flexural strength (FS) and flexural modulus (FM) of a granular lateritic soil collected from the eastern part of India.

2 Materials and Methodology

2.1 Materials

The virgin lateritic soil collected from Jatni, located in the state of Odisha, India was collected for conducting this study. The engineering properties of the soil with respective testing standard is given in Table 1. Grain size distribution of this soil determined as per ASTM D422 (2007) is given in Fig. 1. The cement used for this study was bagged Portland slag cement (PSC), which is generally used for local construction purposes. Two cement percentages (i.e. 5% and 9%) were chosen for this study to achieve the target 7 days UCS value of 3 MPa and 4.5 MPa recommended for cement treated subbase and base respectively by Indian Roads Congress (IRC 2012). The compaction characteristics of the stabilized and un-stabilized soil under modified Proctor energy (ASTM D1557 2012) is presented in Table 2.

2.2 Specimen Preparation and Curing

Prior to the preparation of the specimens, the lateritic soil was air dried and particles larger than 20 mm were removed. The air dried soil was mixed thoroughly with required amount of cement to obtain a uniform mixture. The amount of cement was

Table 1. Engineering properties of soil with testing code

Engineering properties	Value	Testing standard
USCS classification	Clayey sand (SC)	ASTM 2487 (2011)
AASHTO classification	A-2-7	ASTM D3282 (2015)
% finer than 0.075 mm	32.6	ASTM 2487 (2011)
Liquid limit	50	ASTM D4318 (2010)
Plastic limit	23	ASTM D4318 (2010)
Plasticity index	27	ASTM D4318 (2010)
Specific gravity	2.65	ASTM D854 (2014)

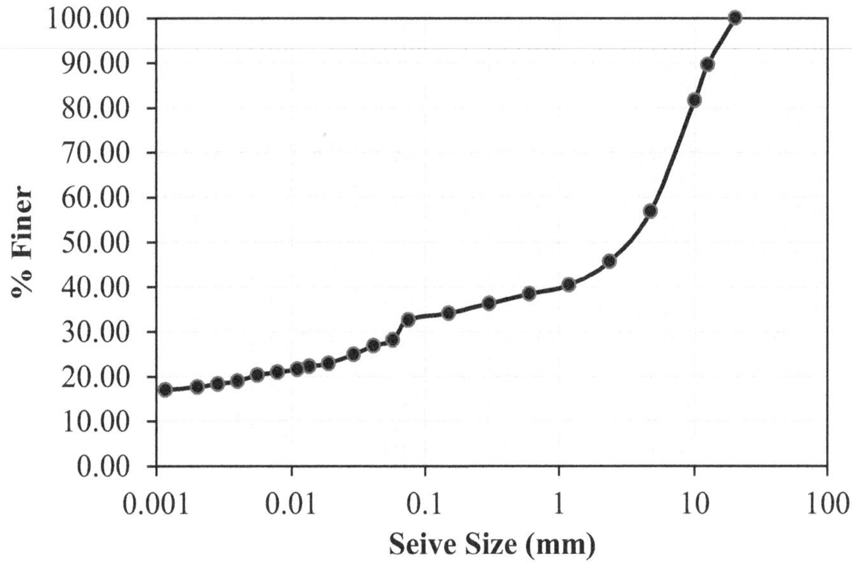

Fig. 1. Particle size distribution of the lateritic soil

Table 2. Optimum water content and maximum density for the cement stabilized materials.

Cement (%)	Optimum water content (%)	Maximum dry density (gm/cc)
0	10.6	1.98
5	11	2.02
9	11.2	2.01

based on the oven dry weight of the soil. An equivalent amount of water representing optimum moisture content was added to the soil-cement mixture and mixed thoroughly. Beam specimens of size 75 mm × 75 mm × 285 mm were prepared for flexural strength and flexural modulus testing and cylindrical specimens of 100 mm diameter and 115 mm height were prepared for UCS test. The specimens were prepared in with appropriate hammer weight, drop height, and compaction effort to achieve the maximum dry density. After compaction, the specimens were removed from the mold and wrapped in polythene. Specimens for UCS test were divided into two sets; one set was cured for 7 days and other for 28 days but the specimens for FS and FM test were cured for only 28 days.

2.3 Testing Program

As there is no test method available to study the effect of W-D cycle on mechanical properties of stabilized material, the standard procedure (ASTM D559 2015) was slightly modified for this study.

Durability Test

One W-D durability test cycle consists of placing the samples inside the potable water (wetting) at 24 °C for 5 h and then putting them inside the oven (drying) at 71 °C for 42 h as shown in Fig. 2. The number of W-D cycles adopted in this study was 5, 10 and 15. After these cycles, samples were tested for mechanical properties.

Fig. 2. Specimens subjected to W-D cycles

Unconfined Compressive Strength (UCS) Test

The UCS tests were performed according to ASTM D1633 (2017) on cylindrical unsoaked samples.

Flexural Strength Tests

The flexural strength (FS) or modulus of rupture (MOR) of the beam samples (size 75 mm × 75 mm × 285 mm) were determined according to ASTM D1635 (2012).

Flexural Modulus Tests

Flexural modulus (FM) is an essential property of the material which is determined from the dynamic flexure test. It is highly dependent on the magnitude of the applied cyclic flexural stress. In absence of a standard procedure to determine the flexural modulus of stabilized materials, the procedure suggested by Yeo et al. (2011) was adopted in this study. The testing arrangement as shown in Fig. 3 was also used for determination of FM. A cyclic haversine load with magnitude of 30% that of FS was applied to the beam specimen at a frequency of 1 Hz. A contact load of 100N was applied to specimen to avoid rocking of the sample during cyclic load. Two Linear Variable Differential Transformer (LVDT) were used to record vertical deflection at midpoint of the beam. A total of 100 load cycle were applied, out of which first 50 cycles were considered for conditioning. The flexural modulus at each loading cycle is determined using Eq. 1. The average modulus value of last 50 cycles was considered as the flexural modulus value of the sample.

$$FM = \frac{23Pl^3}{108bd^3\delta h} \tag{1}$$

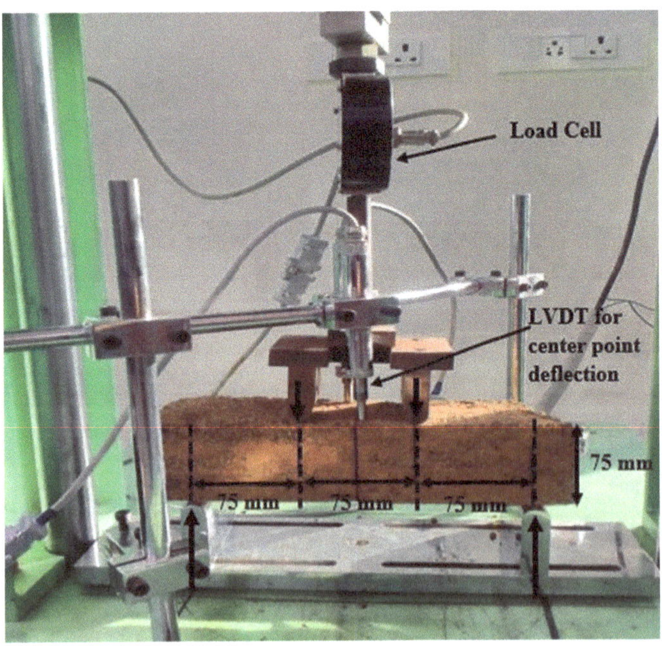

Fig. 3. Testing setup for flexural strength and flexural modulus

Where *FM* = flexural modulus in MPa, *P* = cyclic load in Newton, *l* = span length in mm, *b* = average width of specimen in mm and d = average depth of specimen in mm and δh = peak mid-span displacement in mm.

The stress value is generally chosen to be 30% of the FS. But in this case, as the specimens are subjected to W-D cycles, which will create variation in the FS value, constant value of flexural stresses were chosen to compare the modulus of different specimens. The values were 0.086 MPa for specimens with 5% cement and 0.25 MPa for specimens with 9% cement.

3 Results and Discussion

3.1 Effect of W-D Cycle on UCS

The variations in UCS values for both the cement contents and curing periods are shown in Figs. 4 and 5 respectively. It was observed that, UCS value for all cases increased with increase in number of W-D cycles as shown in Fig. 6. The percentage increase in UCS value for 7 days cured specimens with 5% cement were 45%, 91% and 102% after 5, 10 and 15 W-D cycles respectively. For 28 days cured specimens, this increase was 22%, 55% and 68% respectively.

Similarly, the increase in UCS values of 7 days cured specimens with 9% cement were observed to be 115%, 132% and 137% after 5, 10 and 15 W-D cycles respectively

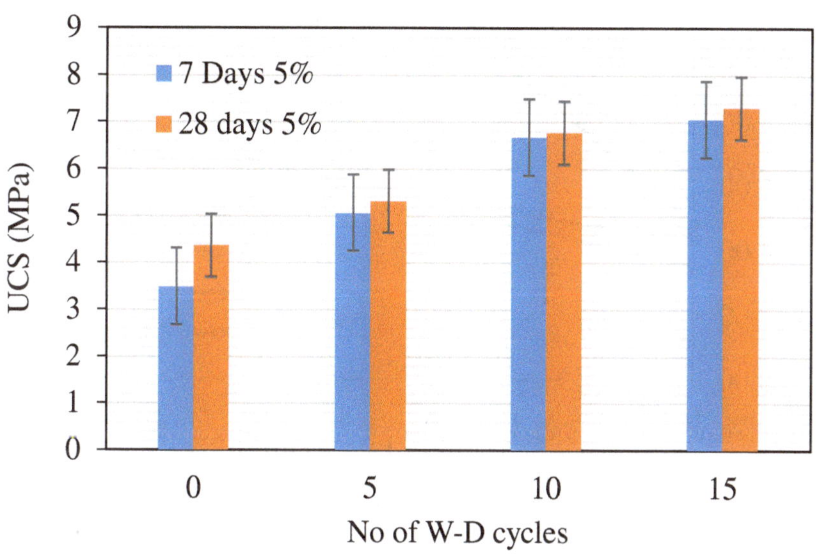

Fig. 4. UCS values at different W-D cycles for 5% cement at different curing periods

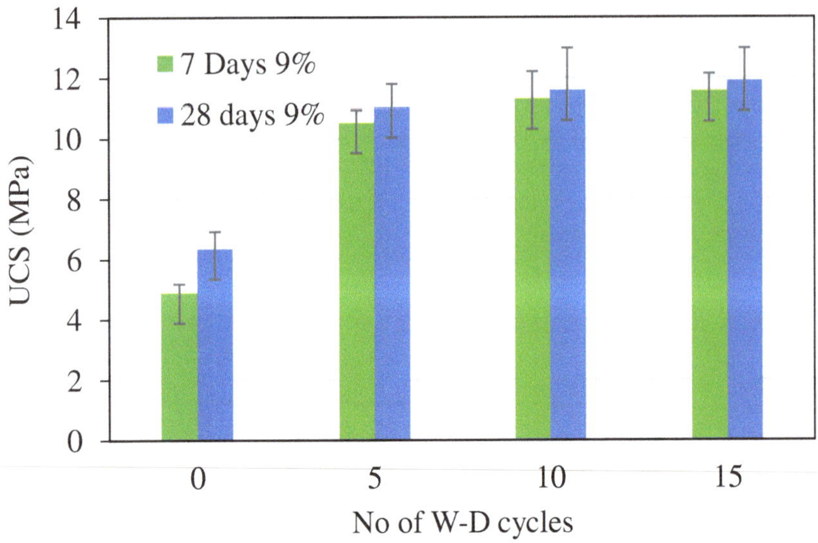

Fig. 5. UCS values at different W-D cycles for 9% cement at different curing periods

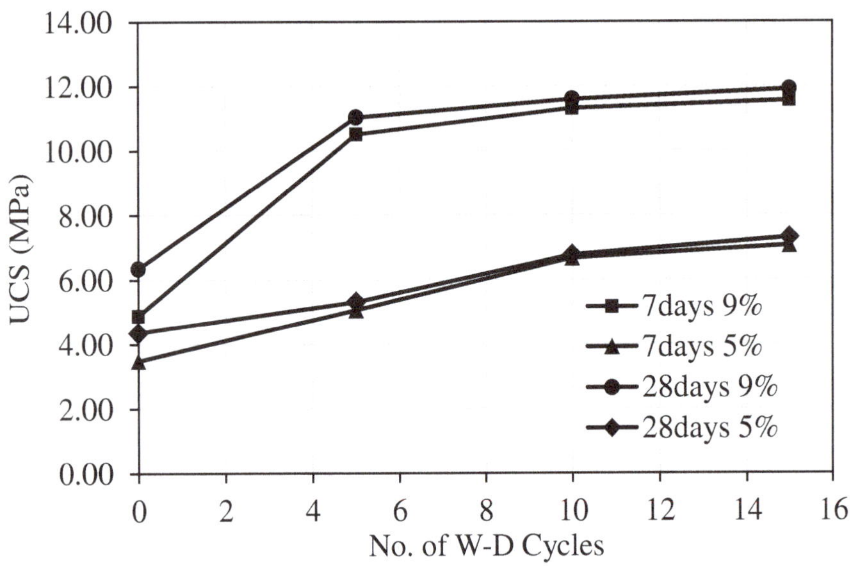

Fig. 6. UCS values of 7-days- and 28-days-cured specimens subjected to W-D cycles

whereas for 28 days cured specimens, these were 74%, 83% and 88%. It may be observed that, the rate of increase slowed down after 5 W-D cycles for all cases. Also, the difference in UCS values between 7 days and 28 days cured specimen reduced after 5 W-D cycles. It may be due to the heating process in W-D cycle which accelerated the curing process for 7 days cured specimens making it at par with 28 days cured specimen in terms of strength gain. A similar trend was also reported by Khoury and Zaman (2002) for class C coal fly ash–stabilized aggregate base.

3.2 Effect of W-D Cycle on Flexural Strength

The variation in FS with number of W-D cycles is shown in Fig. 7. It can be observed that the FS increased with the W-D cycles, which was consistent with the behavior of UCS. The percentage increase in FS after 5, 10 and 15 W-D cycles were approximately 28%, 65%, 96% for 5% cement and 2%, 21% and 29% for 9% cement from specimen not subjected to W-D cycle.

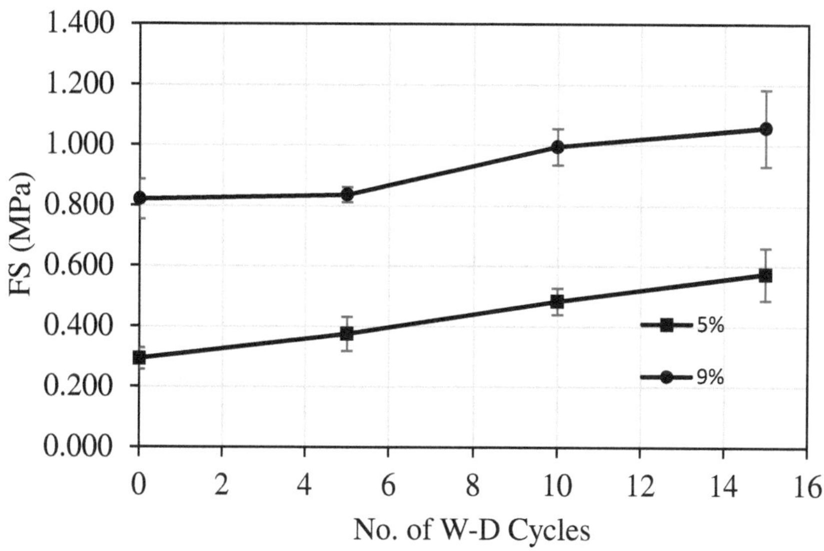

Fig. 7. Variation of FS with W-D cycles

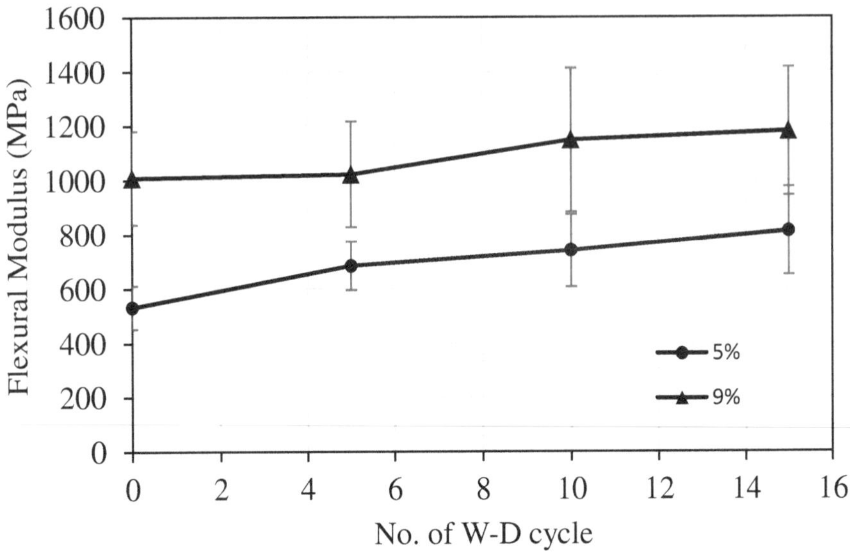

Fig. 8. Variation of FM with W-D cycles

3.3 Effect of W-D Cycle on Flexural Modulus

The increase in FM value due to W-D cycles is shown in the Fig. 8. The percentage increase in FM after 5, 10 and 15 W-D cycles were found to be 29%, 39%, 53% approximately for 5% cement and 1%, 14% and 17% for 9% cement compared to the specimens not subjected to W-D cycle. From these results, it is quite evident that the trends of the FM is quite similar to that of UCS and FS.

4 Conclusions

This study was undertaken to investigate the effect of W-D cycles on the strength and stiffness parameters of cement stabilized granular lateritic soil. The results revealed that with the increase in W-D cycles, strength and stiffness parameters increases. However, the rate of increase after 5 cycles slows down. Curing period was found to have minor effect on UCS after 5 cycles. Similar trends were also observed for FS and FM. This may be attributed to the higher temperature during drying process, which may have accelerated the curing process helping in early strength gain. This increase in strength gain probably negated the effect of wetting. Though the mass loss during W-D process usually increases with increase in number of cycles indicating strength loss of the material, results of this study showed a different trend with increase in stiffness value. Therefore, further studies are needed on other lateritic soils with different cement percentages to verify the conclusion derived from this study.

References

ASTM D854: Standard Test Methods for Specific Gravity of Soil Solids by Water Pycnometer (2014)
ASTM D422-63: Standard Test Method for Particle-Size Analysis of Soils (2007)
ASTM D4318: Standard Test Methods for Liquid Limit, Plastic Limit, and Plasticity Index of Soils (2010)
ASTM D2487: Standard Practice for Classification of Soils for Engineering Purposes (Unified Soil Classification System) (2011)
ASTM D1635/D1635M: Flexural Strength of Soil-Cement Using Simple Beam with Third-Point Loading (2012)
ASTM D1557: Standard Test Methods for Laboratory Compaction Characteristics of Soil Using Modified Effort (2012)
ASTM 3282: Standard Practice for Classification of Soils and Soil-Aggregate Mixtures for Highway Construction Purposes (2015)
ASTM D559/D559M: Standard Test Methods for Wetting and Drying Compacted Soil-Cement Mixtures (2015)
ASTM D 1633: Standard Test Methods for Compressive Strength of Molded Soil-Cement Cylinders (2017)
Bandara, N., Binoy, T.H., Aboujrad, H.S.: Freeze-thaw durability of subgrades stabilized with recycled materials. In: Cold Regions Engineering 2015, pp. 135–145 (2015)
IRC: 37-2012: Guidelines for the design of flexible pavements. The Indian Roads Congress, New Delhi, India
Kalankamary, G.P., Donald, D.T.: Development of a freeze-thaw test for design of soil-cement. Highw. Res. Rec. **36**, 77–96 (1963)
Khoury, N.N., Zaman, M.M.: Effect of wet–dry cycles on class C fly ash aggregate base. Transp. Res. Rec. J. Transp. Res. Board **1787**, 13–21 (2002)
Khoury, N.N., Zaman, M.M.: Durability of stabilized base courses subjected to wet–dry cycles. Int. J. Pavement Eng. **8**(4), 265–276 (2007)
National Lime Association: Technical brief: mixture design and testing procedures for lime stabilized soil (2006)
Portland Cement Association: Soil-Cement Laboratory Handbook (1992)
Scullion, T., Sebesta, S., Harris, J.P., Syed, I.: Evaluating the Performance of Soil-Cement and Cement-Modified Soil for Pavements: A Laboratory Investigation. Portland Cement Association (2005)
Veisi, M., Chittoori, B., Celaya, M., Nazarian, S., Puppala, A.J., Solis, C.: Accelerated stabilization design of subgrade soils. Research Report No. TX FHWA/TX 06/0-5569-1 (2010)
Ventura, D.F.C.: Durability testing of LCB and CTB materials supplied by Caltrans. Technical Memorandum No. TM-UC-PRC-2003-2, California Department of Transportation (2003)
Yeo, Y.S., Jitsangiam, P., Nikraz, H.: Mix design of cementitious base course. In: International Conference on Advances in Geotechnical Engineering, pp. 379–385 (2011)
Zhang, Z., Tao, M.: Durability of cement stabilized low plastic soils. In: Transportation Research Board 85th Annual Meeting Compendium of Papers CD-ROM, Paper No. 06-1255 (2006)

Influence of the Carbonate Content on CPT-Results and Practical Relevance for the Soil Improvement Works on Major Land Reclamation Projects

Robert Thurner[1(✉)], Johannes Reinisch[2], and Franz Tschuchnigg[2]

[1] Keller Grundbau, Vienna, Austria
r.thurner@kellerme.com
[2] Graz University of Technology, Graz, Austria
j.reinisch@alumni.tugraz.at,
franz.tschuchnigg@tugraz.at

Abstract. This paper describes the challenges that arise during quality control of deep vibro-compaction works on land reclamation projects realized in the Middle East. The aim of deep vibro-compaction is to densify the fill to a certain relative density D_r and thereby assure certain minimal criteria related to the soil behavior such as stiffness and resistance against liquefaction. For quality control of the compaction works, the achieved relative density throughout the hydraulic fill can be measured indirectly by performing Cone Penetration Tests (CPT). The targeted relative density is linked to a cone resistance curve calculated with well-known and established correlation methods like Schmertmann [1] or Baldi [2]. However, the use of these empirical correlations in calcareous sand is questionable because they are not suitable to reproduce the characteristics of these sands. Almeida et al. [3] have shown that carbonate sands show significantly lower q_c-values than silica sands under similar conditions which can lead to unattainable target cone resistance values demanded by the client.

In this paper, the Karlsruhe Interpretation Method (KIM) is used as an alternative approach to traditional CPT interpretation methods in sands that contain a high carbonate content. The KIM has shown to be able to accommodate the intrinsic characteristics of calcareous sands and more reliable in establishing a correlation between the cone resistance q_c and the relative density D_r.

1 Introduction

The quality control of the vibro-compaction of reclaimed land is correlated to the density of the fill which is linked to certain minimum requirements of shear strength and stiffness. Densification requirements are translated into technical parameters like the relative density D_r which can be measured directly by the sand replacement method or indirectly with a correlation between the cone resistance q_c obtained from a Cone Penetration Test (CPT) and the relative density D_r.

Schmertmann [1], Baldi [2] and Meier [4] developed such correlations based on calibration chamber tests performed on a wide range of different silica sands.

H. Shehata and H. Poulos (Eds.): GeoMEast 2018, SUCI, pp. 122–131, 2019.
https://doi.org/10.1007/978-3-030-01917-4_10

However, as most of the material used for land reclamation in the UAE consists, at least partially, of calcareous sand, these correlations are not suitable for deriving a compaction requirement.

Calcareous sands consist of varying amounts of skeletal remains of marine organisms such as fragmented seashells and corals and show a high tendency towards grain crushing. This behavior has been identified by Almeida et al. [3] as the reason why, compared to silica sands, calcareous sands produce significantly lower q_c-values in a similar soil state.

An alternative to existing correlations is the semi-empirical Karlsruhe Interpretation Method (KIM). The advantage of the application of the KIM is that the necessary hypoplastic parameters are derived from standard laboratory tests performed on sand samples retrieved from the site, which allows to describe the mechanical behavior of the actual soil on the site.

2 Influence of the Carbonate Content on Soil Properties

Calcareous sands are defined as sands predominately made up of carbonate minerals, most commonly calcite, and are therefore also frequently referred to as carbonate sands. In general, the carbonate sands used for land reclamation projects in the UAE are formed by the accumulation of skeletal remains and fragments of small sea organisms, such as seashells and corals (bioclastic deposits).

The carbonate content of the investigated materials varies between 9.73% and 100% in a database of 11 sand samples. It was found that the carbonate content has a major influence on four basic geotechnical parameters: the limit void ratios e_{min} and e_{max}, the critical friction angle φ_c' and the grain density ρ_s. All three parameters increase with an increasing carbonate content of the sand. It is assumed that this behavior is caused by the angular shape of seashells and their fragments, which form strongly interlocking grain skeletons with a high number of voids. Figure 1 shows the increase of the void ratios (e_{min} and e_{max}) with the carbonate content for the sand samples in the database. The grain density ρ_s is influenced by the mineralogy of the sand. Seashells and other skeletal remains of marine organisms consist of calcite or aragonite with a density of 2.7 g/cm^3 and 2.9 g/cm^3 respectively and therefore cause the grain density to increase as well.

Especially the void ratios and the critical friction angle have a significant influence on the soil's unit weight γ and therefore also on the resulting mean effective stress p_0', which is used in the equation of the KIM. With an increasing carbonate content, γ and p_0' decrease and affect the resulting q_c-requirement calculated with the KIM.

3 CPT Interpretation Methods

The success of the compaction works is measured by comparing the measured cone resistance values in a CPT to a q_c-curve calculated for a certain relative density. Schmertmann [1], Baldi [2] and Meier [4] have proposed correlation methods which allow the calculation of such q_c-curves and assess the in-situ relative density D_r.

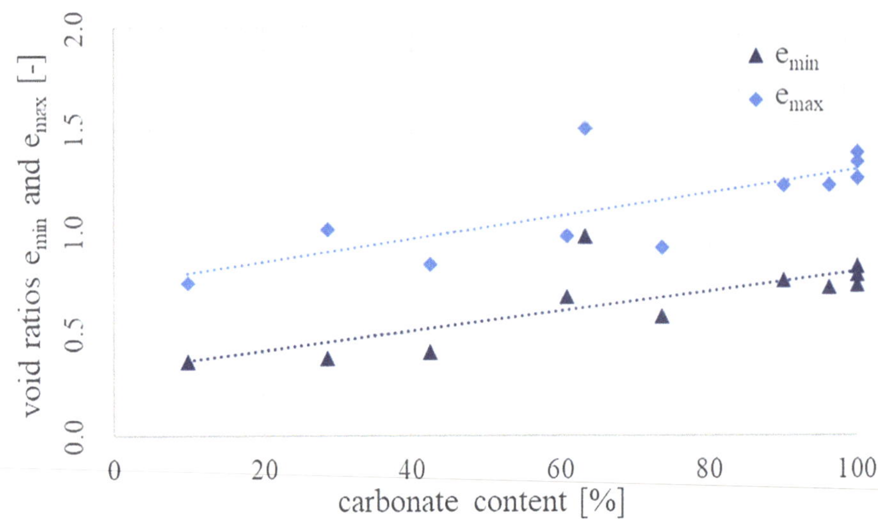

Fig. 1. The influence of the carbonate content on the void ratios e_{min} and e_{max}

These empirical correlation methods are based on the results of calibration chamber tests on silica sands and are commonly used to develop a q_c-target curve corresponding to the relative density D_r demanded by the client.

3.1 Interpretation Method by Schmertmann and Baldi

The method developed by Schmertmann [1] is an empirical relationship between the cone resistance q_c and the relative density D_r which was developed based on the results of 80 calibration chamber tests on silica sands. Equation (1) shows the equation for interpreting CPT measurements:

$$q_c = C_0 * \sigma_v'^{(C_1)} * exp^{(D_r * C_2)} \tag{1}$$

The C_i-constants take into account different material parameters whereas the stress-state is incorporated by the effective vertical stress σ_v'. However, the C_i-values by Schmertmann [1] and Baldi [2] listed in Table 1 do not account for grain crushing or the lower compressibility of calcareous sands which are the main reasons for the differences between the cone resistance in silica and calcareous sands. The C_i-constants developed by Meier [4] (see Table 1) were derived from calibration chamber tests performed on a calcareous sand from Dubai.

Note that in order to use Schmertmann's equation together with the listed C_i-values from Schmertmann and Meier, the value of q_c and σ_v' have to be inserted in kgf/cm^2 (1 kPa = 0.01019716 kgf/cm^2). Baldi's C_0-value of 157 accommodates the conversion of kPa to kgf/cm^2, which allows q_c and σ_v' to be entered directly without conversion.

Table 1. C_i-constants by Schmertmann [1], Baldi [2] and Meier [4]

Correlation	C_0	C_1	C_2
Schmertmann	12.31	0.71	2.91
Baldi	157	0.55	2.41
Meier	14.23	0.67	2.90

4 The Karlsruhe Interpretation Method

The Karlsruhe Interpretation Method (KIM), developed by Cudmani in [5], is a semi-empirical method based on numerical simulations of a spherical cavity expansion. Cudmani established a relation between the limit pressure p_{LS} obtained from spherical cavity expansion calculations and the cone resistance q_c by introducing a so-called shape factor k_q. The shape factor k_q is derived from the results of CPTs performed in a calibration chamber and represents the empirical part of the KIM-equation:

$$q_c = k_q(D_r) * p_{LS}(D_r, p'_0) \tag{2}$$

Equation (2) consists of two parts, the shape factor k_q and the limit pressure p_{LS}. While the empirically determined shape factor k_q is only dependent on the relative density D_r, the limit stress p_{LS} is derived from the solution of a spherical cavity expansion (SCE) problem.

Term k_q is the shape factor and establishes the relationship between the cone resistance q_c and the limit pressure p_{LS}. By analyzing the results of calibration chamber tests on 9 different sands, Cudmani [5] found that the shape factor solely depends on the relative density and can be approximated with:

$$k_q = 1.5 + \frac{5.8 * D_r^2}{D_r^2 + 0.11} \tag{3}$$

In contrast to the interpretation methods presented in the previous paragraphs, which are based on empirically determined parameters, the Karlsruhe Interpretation Method combines an empirical parameter and numerical simulations to calculate the cone resistance corresponding to a certain state of the soil.

4.1 Hypoplasticity and Spherical Cavity Expansion

Cudmani applied the hypoplastic model developed by von Wolffersdorff [7] in the context of the KIM. The hypoplastic equation is calibrated based on the soil found on the specific site and is used to model the spherical cavity expansion (SCE) problem. Von Wolffersdorff's model requires eight parameters for the calibration. These parameters are determined by basic lab-tests performed on disturbed samples taken from the site under investigation, which guarantees that the characteristics of the soil on

Table 2. Input parameters for the hypoplastic constitutive model and the necessary lab tests [7]

Parameter		Lab test
Critical friction angle	φ_c	Angle of repose test
Limit void ratios	e_{i0}	Densest and loosest packing
	e_{c0}	(DIN 18126 [12])
	e_{d0}	
Granulate hardness	h_s	High-pressure oedometer
Exponent	n	(DIN 18135 [13])
Exponent	α	Drained triaxial test
Exponent	β	(DIN 18137-2 [14])

site are properly represented in the simulation. Table 2 lists the necessary parameters and the corresponding laboratory tests. A detailed description of these parameters and the process of their determination can be found e.g. in von Wolffersdor [7], Bauer [8], Herle and Gudehus [9] and Herle [10].

The hypoplastic model by "von Wolffersdorff" is only valid for a so-called simple grain skeleton which presumes that the grains are indestructible and therefore, does not account for abrasion and/or crushing of sand particles. However, Meier [4] demonstrated that the KIM works well for a sand containing breakable grains, if an accurate calibration of the hypoplastic model parameters is assured. To accommodate the lower compressibility of calcareous sands and potential grain crushing, the hypoplastic constitutive equation has to be calibrated for a comparable pressure range as obtained in the actual CPTs. Therefore, the maximum vertical pressure σ'_v in the oedometer test can be as high as 25 MPa.

Parameters h_s, n, α and β are found by curve fitting the results of drained triaxial and high-pressure oedometer tests with the aid of a hypoplasticity element test program. This entails a certain sensitivity which extends into the simulation of the SCE and further on into the resulting q_c-requirement.

The limit pressure p_{LS} is obtained from the numerical solution of a spherical cavity SCE problem. The simulation model is built up of two concentric spheres A and B. Sphere A expands quasi-statically and symmetrically inside of sphere B which represents the adjacent soil (modelled with the calibrated hypoplastic constitutive model). The expansion of sphere A progresses with a constant velocity, sphere B is loaded with the initial mean effective stress p'_0 (which is kept constant during the simulation). With the continuing expansion the stress inside of sphere A and the void ratio of the adjacent material approach limit values that correspond to the critical state of the soil. For a given soil, these limit values depend on the initial state governed by the initial mean effective stress p'_0 and the initial relative density D_r.

Cudmani and Osinov [11] have demonstrated that the limit pressure p_{LS} and the cone resistance q_c depend on the state of the soil in the same manner and increase with an increasing relative density D_r and mean effective stress p'_0. After Cudmani and

Osinov [11] solved the SCE-problem for five different silica sands and various initial conditions, they found that the radial limit pressure p_{LS} as a function of the initial state can be approximated by the equation:

$$p_{LS} = a * p_0'^b \tag{4}$$

$$a = a_1 + \frac{a_2}{a_3 + D_r} \tag{5}$$

$$b = b_1 + \frac{b_2}{b_3 + D_r} \tag{6}$$

The a_i and b_i parameters are determined by curve-fitting the results of a series of solutions of the SCE for different initial conditions with Eqs. (5) and (6). Just like the input parameters for the hypoplastic constitutive model, the KIM parameters also entail a certain sensitivity caused by the curve-fitting. This is one reason why the input parameters for the hypoplastic model and the KIM parameters (and therefore also the resulting q_c-value) might show some differences even though the sands basic properties (i.e. e_{min}, e_{max}, φ_c', ρ_s) are very similar.

5 Influence of the Carbonate Content on KIM Results

The KIM was applied to all 11 samples in the database and to show the influence of the sand's carbonate content on the results of the KIM, the calculated q_c-values for each project against the corresponding carbonate content. Figure 2 depicts the relationship between the carbonate content and the cone resistance q_c and clearly shows that the results decrease with an increasing carbonate content. This proves that the characteristics of calcareous sands can be incorporated in the KIM.

However, for samples with a carbonate content below 50% the resulting values of the cone resistance increase. Based on the studies and the available data it is assumed that the carbonate sand fraction only starts to influence the sand's mechanical behavior once it surpasses a certain threshold of around 50%. Further research is ongoing to investigate this behavior in more detail and additional samples in the database are needed to underpin the findings so far. Additionally, to clarify the connection between the soil properties, the input parameters for the hypoplastic model and the KIM parameters, an extensive sensitivity analysis will be carried out as a next step.

6 Practical Application of the KIM: A Case Study

The investigated project is an extension to an existing reclaimed peninsula in Dubai. The proposed extension involves increasing the area of the peninsula by approximately 170 000 m^2 through an extensive land reclamation operation which required approximately 4 million m^3 of additional dredge material. The dredged sand used as hydraulic fill for this project showed high amounts of seashells and coral lumps and was therefore

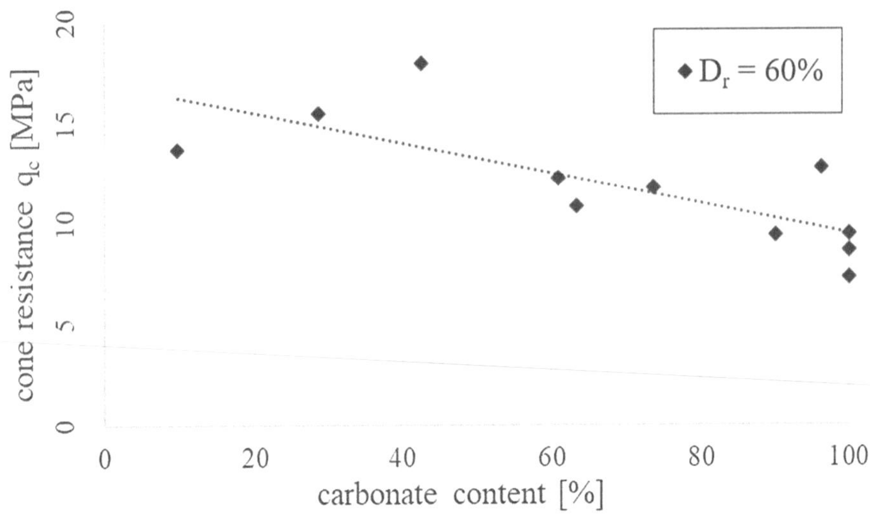

Fig. 2. Relationship between the carbonate content and the cone resistance q_c

expected to exhibit a carbonate content high enough to lead to the known issues with the quality control of the compaction works by means of CPT. It was therefore decided to establish a site-specific correlation between the relative density D_r and the cone resistance q_c.

Two sand samples at two different locations (PLMAZ28 and PLMBC36) were taken and sent to a soil mechanics laboratory. Under the assumption that the dredged sand is more or less homogeneous throughout the whole reclaimed area, it was expected that the hypoplastic and KIM parameters resulting from the lab tests should not differ too much.

The soil mechanics laboratory carried out a series of tests and numerical simulations in order to determine the relevant material and model parameters for the KIM. Before any of the tests were performed, the two samples were sieved and any sand grains larger than $d > 8$ mm were removed due to size limitations of the test apparatuses.

6.1 Soil Properties, Hypoplastic and KIM-Parameters

As the amount of seashells and coral lumps observed (while taking the samples) already suggested, the carbonate content of both samples was determined to be 100% $CaCO_3$. Table 3 sums up the results of the laboratory tests and it can be seen that the differences between the parameters of both sand samples are insignificant.

For the calibration of the respective hypoplastic models of both samples, the laboratory determined the eight necessary parameters listed in Table 4. As can be seen, the differences between the two data-sets are increasing compared to the differences between the basic soil properties listed in Table 3. With a difference of 25.6% and

−28.9% the values of h_s and β of both materials show the biggest variation. It is assumed that this is connected to the interrelation of the parameters h_s, n, α and β.

Table 3. Basic soil properties of the samples PLMAZ28 and PLMBC36

Sample	$CaCO_3$ [%]	e_{min} [−]	e_{max} [−]	φ_c [°]	ρ_s [g/cm³]
PLMAZ28	100	0.74	1.261	36.3	2.791
PLMBC36	100	0.79	1.384	36.5	2.791
Δ [%]	0.00	6.76	9.75	0.55	0.00

After the hypoplastic constitutive model was calibrated for both samples, a series of numerical simulations of the SCE were carried out with a finite difference code. By curve-fitting the results of these simulations, the a_i- and b_i-parameters listed in Table 5 were determined. In comparison to the differences between the basic soil- (Table 3) and hypoplastic parameters (Table 4), the differences between the a_i and b_i values of the two materials increase significantly. Since the two sand samples are practically identical, it is assumed that the discrepancies originate from the sensitivity of the curve-fitting process in the determination of the hypoplastic and the KIM-parameters.

Table 4. Hypoplastic parameters of the samples PLMAZ28 and PLMBC36

Sample	φ_c [°]	e_{d0} [−]	e_{c0} [−]	e_{i0} [−]	h_s [MPa]	n [−]	α [−]	β [−]
PLMAZ28	36.3	0.74	1.261	1.450	39	0.525	0.050	1.97
PLMBC36	36.5	0.79	1.384	1.592	49	0.480	0.045	1.40
Δ [%]	0.55	6.76	9.75	9.79	25.64	−8.57	10.00	28.93

6.2 Results of the KIM for the Investigates Samples

After the laboratory tests, evaluations and SCE simulations were completed and all necessary parameters have been determined, the cone resistance for any desired relative density and depth can be calculated. Figure 3 shows the q_c-curves computed with the values of the PLMAZ28 and the PLMBC36 sample down to a depth of 10 m and a relative density of D_r = 60%.

Table 5. KIM parameters of the samples PLMAZ28 and PLMBC36

Sample	a_1	a_2	a_3	b_1	b_2	b_3
PLMAZ28	1.666	−6.152	−1.597	0.835	0.073	−1.395
PLMBC36	2.550	−4.477	−1.544	0.841	0.060	−1.460
Δ [%]	53.36	−27.23	−3.32	0.72	−17.81	4.66

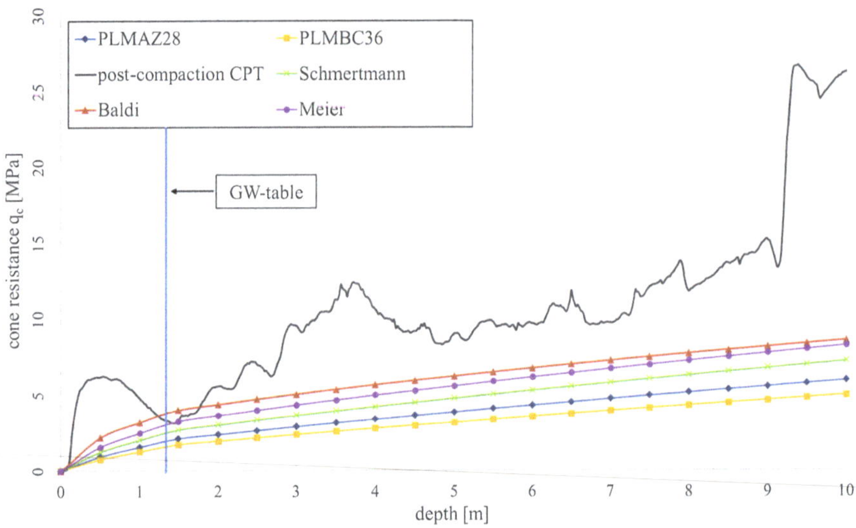

Fig. 3. Cone resistance curves calculated with the KIM for both samples from the investigated project for a relative density of $D_r = 60\%$

The cone resistance curves calculated with the KIM for the samples from PLMAZ28 and PLMBC36 show how the results of the KIM vary even though, as shown in Table 3, the two samples are more or less identical. At a depth of 10 m, the q_c-values of the two samples are $q_c = 5.96$ MPa for the PLMAZ28 sample and $q_c = 4.97$ MPa for the PLMBC36 sample, differing by 0.99 MPa.

7 Conclusion

The findings presented in this paper indicate that the use of the Karlsruhe Interpretation Method is beneficial for the contractor if the sand present at the site is predominately of calcareous composition. In general, it was found that the application of the KIM leads to low cone resistance requirements in sands with a high carbonate content. Another advantage is that the KIM-parameters are derived for every site and therefore produce a site-specific q_c-requirement for every project taking the characteristics of the material present on site into account.

The results of the KIM display the actual material behavior. This is a strong argument against traditional CPT interpretation methods which are based on calibration chamber tests on silica sands and do not incorporate the specific characteristics of calcareous sands. However, further investigations and studies are ongoing to study the connection between the soil properties and the hypoplastic and KIM parameters in detail.

Acknowledgments. The authors of this paper would like to acknowledge Dr. Thomas Meier, Team Manager of Finite Elements at Baugrund Dresden, for his continuous constructive feedback and advice about the KIM. Dr. Meier shared invaluable insights toward the topic of determination of the parameters necessary for the KIM and its practical application.

References

1. Schmertmann, J.H.: An Updated Correlation Between Relative Density Dr and Fugro-Type Electric Cone Bearing, qc. Wicksburg, Mississippi (1976)
2. Baldi, G., Bellotti, R., Ghionna, N., Jamiolkowski, M., Pasqualini, E.: Interpretation of CPTs and CPTUs, 2nd part: drained penetration of sands. In: 4th International Geo-technical Seminar, Field Instrumentation and In-Situ Measurements, pp. 143–155 (1986)
3. Almeida, M.S.S., Jamiolkowski, M., Peterson, R.W.: Preliminary result of CPT tests in calcareous Quiou sand. In: Proceedings of the First International Symposium on Calibration Chamber Testing, pp. 41–53 (1991)
4. Meier, T.: Application of Hypoplastic and Viscohypoplastic Constitutive Models for Geotechnical Problems. Universität Fridericiana in Karlsruhe (2007)
5. Cudmani, R.: Statische, alternierende und dynamische Penetration in nichtbindigen Böden. Universität Fridericiana in Karlsruhe (2000)
6. Kolymbas, D.: A rate-dependent constitutive equation for soils. Mech. Res. Commun. **4**(6), 367–372 (1977)
7. von Wolffersdorff, P.A.: Hypoplastic relation for granular materials with a predefined limit state surface. Mech. Cohesive Frict. Mater. **1**(3), 251–271 (1996)
8. Bauer, E.: Calibration of a comprehensive hypoplastic model for granular materials. Soils Found. **36**, 13–26 (1996)
9. Herle, I., Gudehus, G.: Determination of parameters of a hypoplastic constitutive model from properties of grain assemblies. Mech. Cohesive Frict. Mater. **4**(5), 461–486 (1999)
10. Herle, I.: Granulometric limits of hypoplastic models. TASK Q. Sci. Bull. Acad. Comput. Cent. Gdansk **4**(3), 389–408 (2000)
11. Osinov, V.A., Cudmani, R.: Theoretical investigation of the cavity expansion problem based on a hypoplasticity model. Int. J. Numer. Anal. Methods Geomech. **25**(5), 473–495 (2001)
12. Deutsches Institut für Normung e. V.: DIN 18126:1996-11, Soil investigation and testing - Determination of density of non-cohesive soils for maximum and minimum compactness (1996)
13. Deutsches Institut für Normung e. V.: DIN 18135:2012-04, Soil investigation and testing - Oedometer consolidation test (2012)
14. Deutsches Institut für Normung e. V.: DIN 18137-2:2011-04, Soil investigation and testing - Determination of shear strength - Part 2: Triaxial test (2011)

Strength and Micro Structural Behavior of Lime Stabilized Dredged Soil

Obaid Qadir Jan[1(✉)] and Bashir Ahmed Mir[2]

[1] National Institute of Technology, Srinagar, Jammu and Kashmir, India
Obaidqadir193@gmail.com
[2] Department of Civil Engineering, National Institute of Technology, Srinagar, Jammu and Kashmir, India

Abstract. Dredged soil deposits are generally soft soil deposits, possessing low bearing capacity and high compressibility, generated by the dredging of a river. Using such soil deposits in its natural condition is not recommended. Concern over environmental effects of dredging, disposal and the increasing unavailability of suitable disposal sites has put pressure for characterization of this material. This material can be a valuable resource for many practical purposes, such as fill material, sub-grade construction, reclamation, landscaping etc. In the present investigation, various soil samples from three different locations of flood spill channel were collected. Soil tests like gradation, specific gravity, consistency indexes, light compaction, unconfined compressive strength, and direct shear tests were performed. The results of the three samples were compared and the weakest soil sample was selected for lime treatment. The mechanical properties of the dredged soil were improved using hydrated lime in varying percentages of 4%, 8%, 12% and 16%. Consistency limits, direct shear parameters and compaction characteristics of treated soil samples were studied. The unconfined compressive strength increased significantly with increase in lime content, as well as curing period, however, the optimum value was observed at 8% lime content, beyond which the strength decreased. The SEM micrographs of the soil samples revealed that the structure changed from deflocculated form to flocculated form on addition of lime. Dispersive tendencies are inhibited and several large flocs indicating strong inter particle bonding can be seen. The cementitious compounds formed due to pozzolanic reactions filled the void spaces which improved the strength of the soil.

Keywords: Dredged soil · Floods · Lime · Mechanical properties
SEM · Solid waste

1 Introduction

Dredged soil deposits are soft soils, exhibiting low bearing capacity and high compressibility and are a challenge to geotechnical engineers, as both strength and serviceability requirements may not be satisfied [1, 2]. The low strength and stiffness of such soils results in bearing capacity failures and impermissible settlements causing severe damage to buildings and foundations [3]. Therefore, using such soil in its natural condition is not recommended. Dredged soil is one such soft soil deposit of solid waste,

H. Shehata and H. Poulos (Eds.): GeoMEast 2018, SUCI, pp. 132–153, 2019.
https://doi.org/10.1007/978-3-030-01917-4_11

generated by dredging of a river [4]. Dredged sediment is one of the potential sources of nutrients which are a concern for eutrophication of surface waters [5]. Sediment is a result of water or wind erosion, which disintegrates soil into various components of sand, silt, clay and organic matter. These products are carried into a water body where the particles settle down at the bottom [6].

Dredging is an activity of removing the sediments from the bottom portion of the rivers, lakes and sea beds [7]. Dredging is one of the least known and popularized earth and rock moving methods, but, it is certainly one of the most widespread and consequential ones. Dredging is an essential component of the sound economic management of maritime infrastructure, besides being an environmental necessity in the areas impacted by highly contaminated sediment [8]. The dredged material contains mixture of contaminants like heavy metals, oils and pesticides which makes its disposal problematic for environmental reasons [9]. The occurrence of such contaminants is attributed to the drainage of waste water from municipal and industrial units, [10–12] anthropogenic activities, [13, 14] agricultural runoff from adjacent grounds and atmospheric deposition [15, 16]. The pollution of aquatic environment by heavy metals is a grave concern due to their high toxicity, persistence and abundance [12–17]. Some benthic organisms, flora and fauna, gather heavy metals within their bodies that reach toxic levels. They later enter the human food chain causing a variety of different health problems [17]. When the dredged material dries out and becomes oxidized, the potential exists for contaminants to become soluble, mobile and bio available [18].

Leaching of heavy metals can affect the quality of surface waters and ground waters due to run off and percolation [19]. Kelly [20] found vertical movement of Zn and Cd through soils with a low cation exchange capacity (CEC) into the groundwater. The dredged soil has typical characteristics of the fine grained soil type and generally consists of clay and silt sized particles [21]. The main challenge is to find economical and environmentally sustainable means to solidify and stabilize the solids, so that, they can be used for structural fill or soil cover [22].

This study deals with soft soil deposits of solid waste generated by dredging the flood spill channel of river Jhelum in Srinagar. Due to the deposition of sediment and silt carried by the river during its normal course and floods, the carrying capacity of the river reduced drastically, prompting dredging of the river bed to restore its capacity. Dredging of the river bed and flood channel generates dredged soil in large quantity posing serious health and environmental problems. The harmful effects of dredging on the environment, disposal and scarcity of proper disposal sites, has put pressure for characterization of this material.

Dredged soil is now being regarded as a resource rather than a 'spoil' or 'waste'. Its mineralogy and Geotechnical properties qualify it for use in the manufacture of high value, beneficial use products [23, 24]. Dredged material is an under-utilized resource that can be used for beneficial purposes once physical, engineering, chemical or biological properties have been determined. It may be used as a fill material, in sub-grade construction, reclamation, landscaping, agriculture, covers for landfills, constructing wetlands for water quality improvement, bank stabilization, creation of islands, wildlife habitat wetlands, and amongst others [2]. The range of beneficial applications for dredged soil is diverse, being limited only by the insight of the designer. Beneficial use of dredged material is an integral and necessary part of dredge material management process [24, 25]. A number of techniques are available for dewatering and improving engineering properties of dredged material and other soft soils [26, 27].

Stabilization of clay soils using lime is a well tried and tested ground improvement technique, since the times of ancient China, Mesopotamia, Egypt [28]. Greek and Roman civilizations have reported use of soil lime mixtures [29]. Lime stabilization can be used to either modify or stabilize the clays [30]. Today stabilization of clay soil by addition of lime is a technique widely used throughout the world to improve its use in construction. Lime stabilization of soils has been successfully applied in highway works and is achieved with calcium oxide (quicklime) or calcium hydroxide (hydrated lime) [31]. It is frequently used in embankments, unstable slopes, rail road and airport construction, backfills, bridge abutments, retaining walls, canal linings and lime piles [32, 33]. The stabilizing effect depends on the reaction between lime and the clay minerals [34]. The addition of lime to the soil modifies its physical properties, transforms the soil into a stabilized mass which increases its strength and durability [35].

Generally, the lime-stabilized soils are compacted to achieve the double benefit of mechanical and chemical stabilizations [36]. Lime works best for clayey soils, especially those with moderate to high plasticity index (PI > 15) [37]. Croft [38] observed that addition of lime decreases the liquid limit, plasticity index and maximum dry density of the soil, and increases its optimum water content, shrinkage limit and strength. Lime has an optimal dosage value for maximum strength gain which depends on the soil type and clay mineralogy [39]. Bell [34] suggested an optimum percentage of lime needed for maximum modification of the soil lies between 1 and 3% lime by weight. Further, addition of lime does not change the plastic limit, but increases strength. However, other researchers have reported the optimum percentage of lime between 2 and 8% in soil stabilization [40].

The amount of lime required for modification and stabilization of in-situ soil is termed as optimum lime content (OLC). It is defined as the amount of lime which can provide sufficient amount of free calcium required for short term improvement of soil in terms of its workability and plasticity, as well as long term enhancement of its strength, stiffness and durability [41]. Lime reacts with clayey soils in two major ways, one in the short-term and second in the long-term. The short-term process has an immediate effect on the properties of soils, which involves ion exchange between calcium ions from lime and cations near the clay particle surface. This is possible if the calcium ions have a higher charge or higher density than the cations near the clay particle surface. This ion exchange is quite useful, since it is likely to transform the soil from a weak dispersed structure to a strong flocculated structure. It is this process of flocculation which is essentially responsible for modification of engineering properties of clay soils mixed with lime [30]. The long-term pozzolanic reactions begin as an increase in hydroxyl ions from the lime causes an increase in the pH of soil water, which dissolve the silicate and alumina sheets of the clay. As the silica and/or alumina are released, they combine with the calcium to form calcium silicate hydrates and/or calcium alumina hydrates, which cement the clay particles together [42].

This paper, further studies the micrographic changes in the soil particle orientation on the addition of lime and variation in curing period. The microstructure of lime treated clays was studied by Locat et al. amongst others [43]. The present study analyzes the mechanical behavior of lime treated dredged soil, with the results that an optimum lime percentage gives the maximum compressive strength which is confirmed by SEM micrographs. Thus, the main aim of this paper is in establishing that dredged material can be used as a valuable resource for various beneficial applications in the field, besides reducing the environmental concerns caused by its large scale dumping, thereby, helping in sustainable development of the environment.

2 Environmental Issues and Concerns

There are two types of problems associated with generation of huge quantities of dredged soil

1. Environmental problems
2. Land disposal problems

The dredging process may lead to the effects due to the excavation of sediments at the bed, loss of material during transport to the surface, overflow from the dredger at the time of loading and loss of material from the dredger or from the pipelines during transport [44]. Dredged material has sufficient potential to impact the environment adversely. Carriage water and hydrostatic water from various hydraulic and mechanical dredging processes has the capacity to pollute the surface water when discharged to waters of the state. The management of dredged material also has the potential to affect ground water through on-land management methods, if not managed in consideration of environmental risk factors. Dredged material is attracting the attention of researchers/environmentalists because the quantity of material dredged each year continues to increase. Increasing urbanization and industrial development near waterways and ports has made it difficult to locate new sites for dredged material disposal in many areas. The cost of dredged material disposal has increased rapidly with greater distances from the dredging site to the disposal site and within environmental controls [45]. Due to dredging of the flood spill channel of river Jhelum in Srinagar, there is accumulation of heaps of dredged material along the banks of the river. This causes not only the serious health and environmental problems, but also damages the soothing aura around the water body. Thus, there is a dire need for characterization of this material so that possible environmental risk hazards posed by dredged material can be mitigated.

3 Materials and Methods

3.1 Materials

In this study, in-situ undisturbed and disturbed soil samples of dredged material from three different locations of the flood spill channel were collected. The samples were sealed and transported with utmost precaution for studying their various properties and behavior on application of lime. Commercially available hydrated lime was used as an additive for stabilization of the dredged material. Hydrated lime is calcium hydroxide, $Ca(OH)_2$, produced by reacting quicklime (CaO) with sufficient water to form a white powder and the process is known as slaking. Calcium oxide (quick lime) may be more effective in some cases, however, it can cause corrosion of the equipment and may cause severe skin burns to the personnel [46].

The addition of lime to clay soils results in immediate improvement in strength and plasticity followed by the long term reactions leading to the formation of compounds similar to those found in concrete. Applications of lime stabilization include mix- in-place techniques for both sub grade and bulk fill equipment, contaminated land and slope stabilization.

3.1.1 Testing Methodology and Experimental Program

The disturbed soil samples were subjected to various soil tests like gradation, specific gravity, consistency limit tests, shear strength tests and light compaction tests using relevant standard procedures [47, 48]. To check for the presence of any organic content, consistency limits [49] were performed on both air dried and oven dried soil samples. Unconfined compressive strength [50] and direct shear tests [51] were conducted on in-situ undisturbed soil samples to establish shear strength parameters. After geotechnical evaluation of the physical and mechanical properties of the dredged soil, the weakest soil sample was selected for lime stabilization. The physical properties of the selected soil sample are given in Table 1.

Test specimens were prepared with dredged soil using a range of lime from 4 to 16% (with 4% increment by dry weight of soil) at $0.95\gamma_{dmax}$ and optimum moisture content [1, 52]. The lime treated specimens were subjected to consistency limit tests, direct shear tests and unconfined compressive strength tests at immediate, 7 days and 28 days of curing period [53–55].

4 Results and Discussions

4.1 Physical and Engineering Properties of Untreated Dredged Soil

Dredged soils exhibit different range of colors with a single sample showing a whole range of colors merging more or less noticeably into one another forming different patterns [56]. The dredged soil in this study was brownish in color with specific gravity varying between 2.67 and 2.69. The particle size distribution curve depicts the nature of size gradation, range of particle sizes, and a comparison of different soils. The particle size analysis is used in the classification of coarse grained soil. It provides an index to the strength and load bearing capacity of soils, to know the susceptibility of a soil to frost action and in the design of drainage filters [1, 56]. Generally, a well graded soil has higher strength compared to the uniformly or poorly graded soil. The gradation curve in the Fig. 1 illustrates that the soil of all the three sites contains maximum silt content with some percentage of clay.

Consistency limits such as liquid limit, plastic limit and shrinkage limit are widely used in soil mechanics. The values of liquid limit and plastic limit are very useful in the classification of soils and can be used to assess the amount of dewatering needed before a dredged material can be handled and processed. They also provide an overall idea for the engineering properties of the soils. Consistency limit tests were conducted as per

Table 1. Physical properties of soil

S. No.	Property	Value		
		Site 1	Site 2	Site 3
1.	Natural water content, w_n (%)	32	33.2	31.4
2.	Permeability coefficient, k (m/s)	1.4×10^{-8}	3.3×10^{-8}	2.2×10^{-7}
3.	Field dry unit weight, γ_d (kN/m^3)	14.5	13.8	13.9
4.	Specific gravity, G	2.67	2.68	2.69
5.	Clay (%)	13	16.9	10.2
6.	Silt (%)	77	78.4	84.8
7.	Sand (%)	9.7	3.9	3.9
8.	Gravel (%)	0.3	0.80	1.1
9.	Liquid limit, LL (%)	38	39	43
10.	Plastic limit, PL (%)	28	28	29
11.	Shrinkage limit, SL (%)	16	17	21
12.	Plasticity index, PI (%)	10	11	14
13.	Classification (ISC)	MI	MI	MI
14.	Consistency index, I_c	0.6	0.5	0.8
15.	Liquidity index, I_l	0.4	0.5	0.2
16.	In-situ UCS, q_u (kN/m^2)	83.6	31.8	44.2
17.	In-situ cohesion, c_u (kN/m^2)	26.6	5.5	20.7
18.	Angle of internal friction, ϕ_u (Deg)	20.4	13.9	16.1
19.	Optimum moisture content, OMC (%)	22	26	21.5
20.	Max. dry unit weight, (kN/m^3)	16.5	15.4	16.3
21.	Un-soaked CBR, (%)	3.9	2.4	3.2
22.	Soaked CBR, (%)	3.1	2.3	2.6

relevant standard testing procedures [49]. Based on test results, the dredged soil can be classified as clayey silt of medium plasticity as per Indian system of soil classification (ISC) system [47]. The liquid limit and plasticity index of the untreated dredged soil lies in the range of 38–43% and 10–14% respectively. From Fig. 2, it can be seen that the rate of loss of shear strength of site-2 is higher than site-1 and site-3 respectively.

The compaction curves of the untreated dredged soil as shown in Fig. 3 were obtained using the standard Proctor compaction test method [57]. The maximum dry unit weight lies in a narrow range of 15.4–16.5 kN/m^3 and optimum moisture content between 21–26% respectively. The undisturbed in-situ soil samples were tested in an unconfined compression testing machine [50]. The test results gave unconfined compressive strength values of 83 kPa, 31 kPa and 44 kPa for site 1, site 2 and site 3 respectively (Fig. 4). The test values reveal that the soil in its in-situ condition is in soft state.

Direct shear tests were conducted on direct shear testing machine under un-drained conditions as per standard testing procedures [51]. From Fig. 5, it is deduced that cohesion (c) varies in the range of 5–27 kPa and angle of internal friction (ϕ) varies in the range of 13–21° respectively, depicting the loose condition of the soil. The test results on the untreated dredged soil demonstrate that the dredged soil in natural state is not suitable as an engineered construction material.

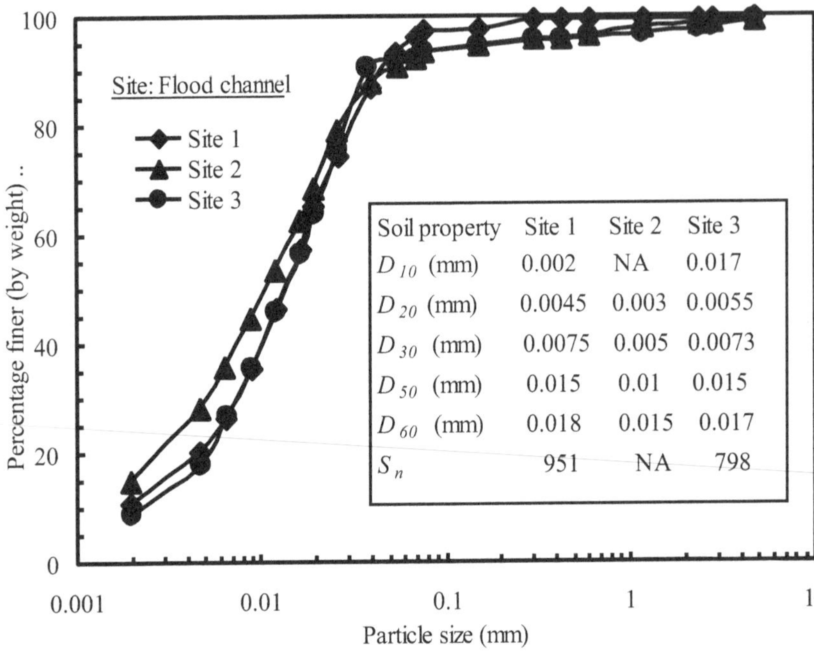

Fig. 1. Particle size distribution curve of dredged soil

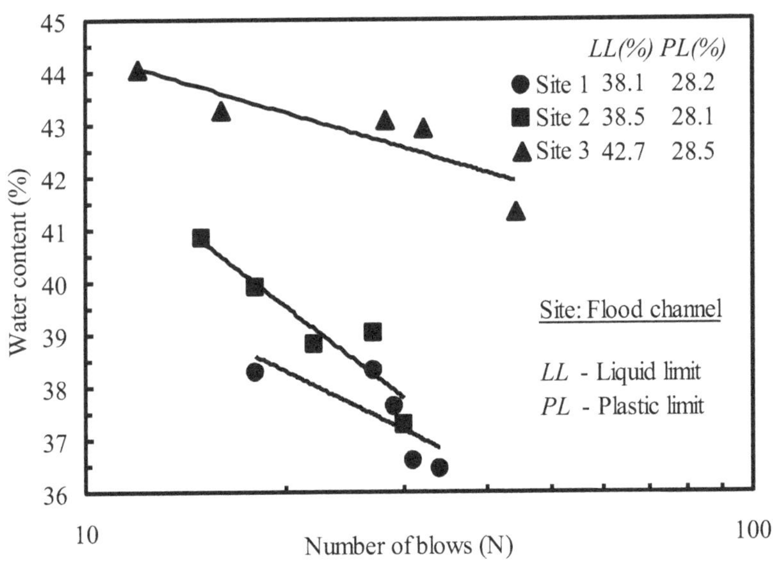

Fig. 2. Flow curves of untreated dredged soil

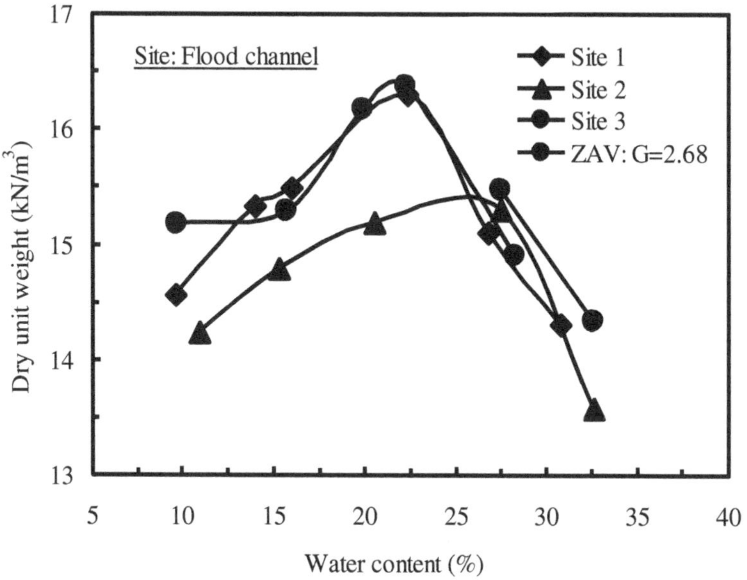

Fig. 3. Compaction curves of untreated dredged soil

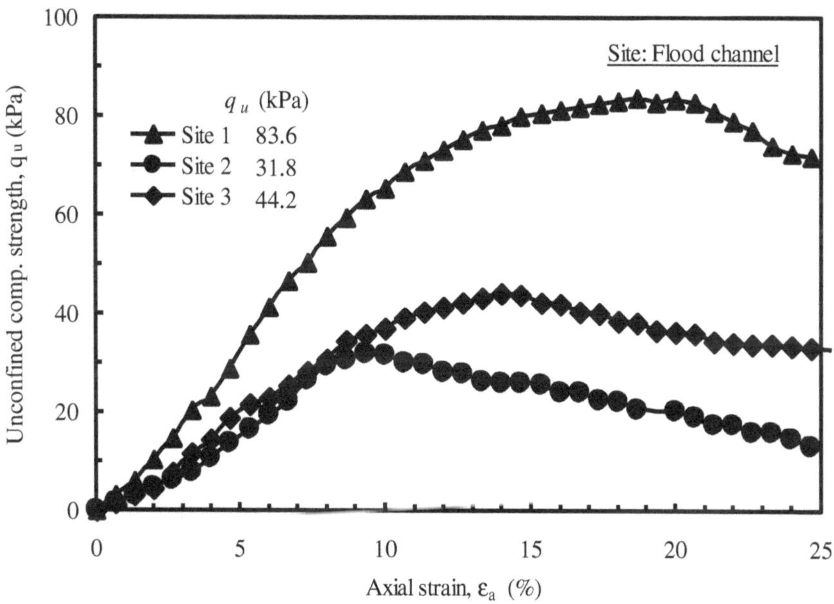

Fig. 4. Unconfined compression test of untreated dredged soil

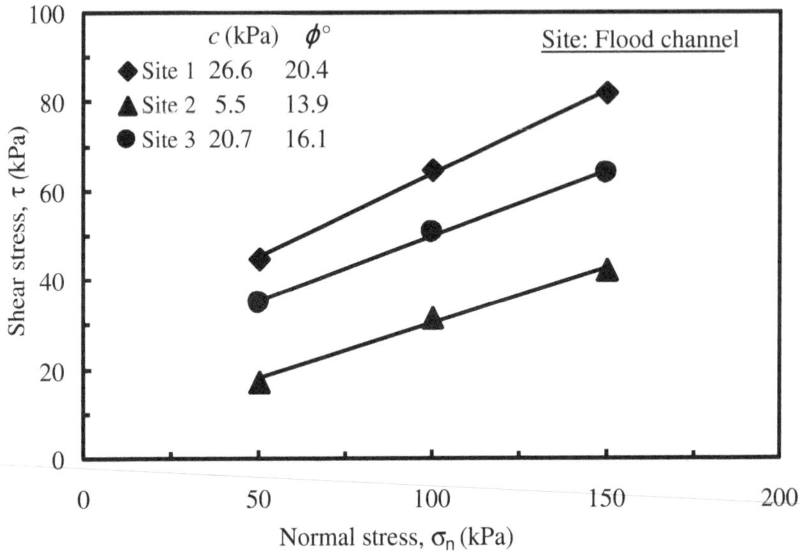

Fig. 5. Mohr Coloumb failure envelope of untreated dredged soil

4.2 Physical and Engineering Properties of Lime Stabilized Dredged Soil

4.2.1 Effect of Lime on Index Properties

The addition of lime to the dredged soil increased both liquid limit and plastic limit of the soil hence reducing its plasticity index. This behavior was observed up to 12% lime content, after which the liquid limit and plastic limit decreased. The minimum value of flow index was seen at 8% lime content, depicting stable nature of the treated specimens at this percentage (Fig. 6). Many other researchers have reported increase in liquid limit

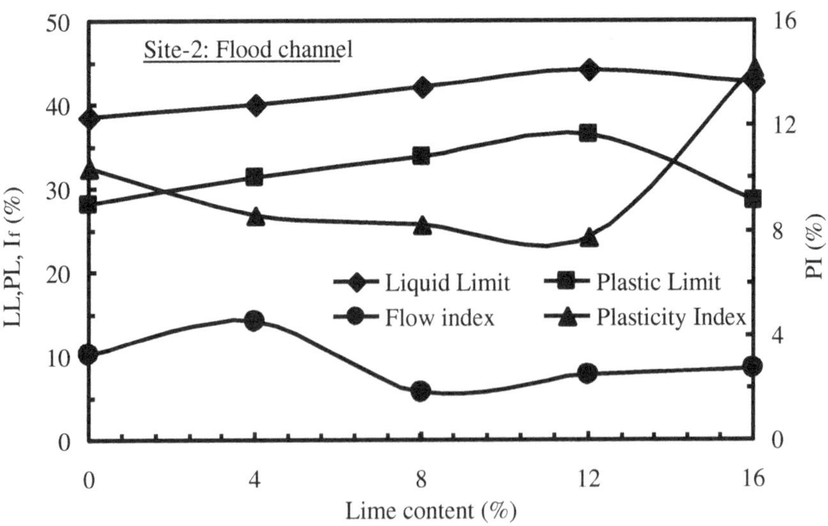

Fig. 6. Effect of lime on index properties of dredged soil

and plastic limit of soil on addition of lime with a decrease in its plasticity index. The first increment of lime to the soil results in sharp increase in the plastic limit and liquid limit which becomes more or less steady with further addition of lime [35, 58].

4.2.2 Effect of Lime on Compaction Characteristics

Compaction is the process of increasing the density of the soil by expulsion of air from the voids on imparting energy to the soil [59]. Compaction is one of the basic and economical methods of soil stabilization. It improves the engineering properties of the foundation soil so that required shear strength is obtained [60]. Generally, a high level of compaction of soil enhances the geotechnical parameters of the soil, so that achieving the desired degree of relative compaction necessary to meet specified or desired properties of soil becomes imperative [61].

In this study, standard Proctor compaction tests were conducted on the dredged soil stabilized with lime at varying percentages of 4%, 8%, 12% and 16% (by dry unit weight of soil) and test results are presented in a plot of dry unit weight versus water content as shown in Fig. 7. From Fig. 7, it is seen that lime treatment flattens the compaction curve, achieving the prescribed density over a much wider range of moisture contents, so that relaxed moisture control specifications are possible. This is attributed to the decrease in thickness of diffuse double layer owing to increased cation concentration on addition of lime, making the soil more friable. The maximum dry unit weight of the stabilized samples increases, and optimum moisture content decreases up to 8% of lime content, beyond which the behavior becomes opposite. Ismail [35] has also reported decrease in maximum dry unit weight and increase in optimum moisture content of lime stabilized soil samples. Hussain and Dash [62] have also come up with similar findings.

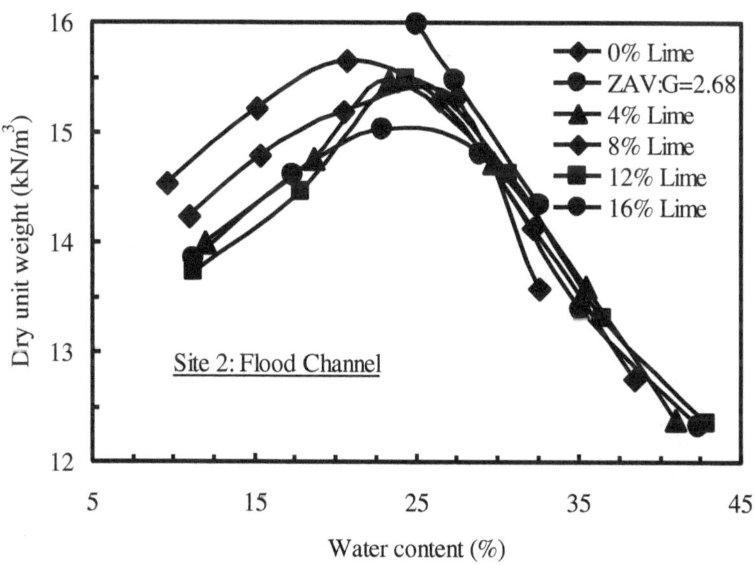

Fig. 7. Lime stabilized compaction curves of dredged soil

4.2.3 Effect of Lime on Strength Characteristics

Strength characteristics are essential parameters that determine the feasibility of engineering application of a soil. Unconfined compression test is the simplest and quickest method to determine the shear strength of cohesive soils. The treated soil specimens were prepared, compacted at $0.95\gamma_{dmax}$ and optimum moisture content under standard procedure. The samples were tested immediately, after 7 days and after 28 days of curing period as shown in Fig. 8. The soil consistency is improved and the unconfined compressive strength (UCS) of dredged soil increases rapidly. Figures 8 and 9 show that the UCS of dredged soil gets enhanced when lime stabilized dredged soil is cured for a period of 7 days and 28 days respectively. It is seen that with increase in lime content, UCS and effective cohesion of soil also increases.

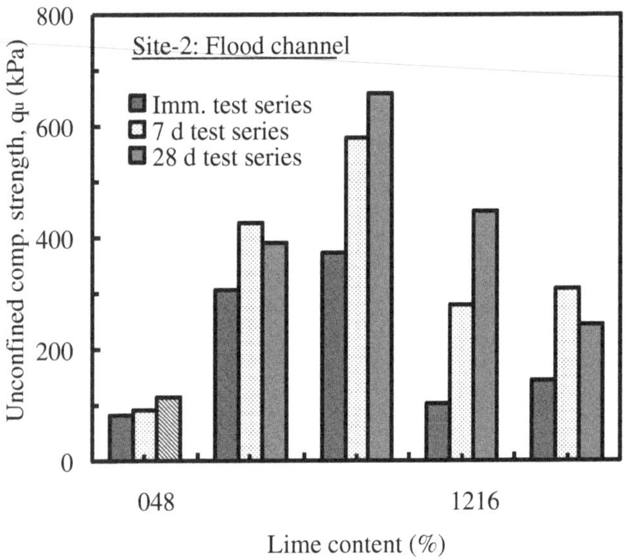

Fig. 8. Strength variation for soil specimens with curing time

The test results reveal that the addition of lime has significant effect on the strength gain of the dredged soil. The optimum lime content obtained which gives maximum unconfined compressive strength is at 8%, (Fig. 9) beyond which the strength decreases. When lime is added to the soil, Ca^{2+} ions migrate to the surface of the clay particles which develops electrical force of attraction between the Ca^{2+} ions and negatively charged clay particles. This leads to the formation of flocs and increase in strength of the soil. When the optimum value of lime content is reached, too many Ca^{2+} ions become available resulting in repulsive force between the particles leading to deflocculation and splitting of the clay particles. This results in development of cracks at smaller loads and early strain rates, hence reducing the unconfined compressive strength of the soil. Amongst various factors, curing is one of the most important factors affecting the strength of lime stabilized clay [63]. The strength increases rapidly

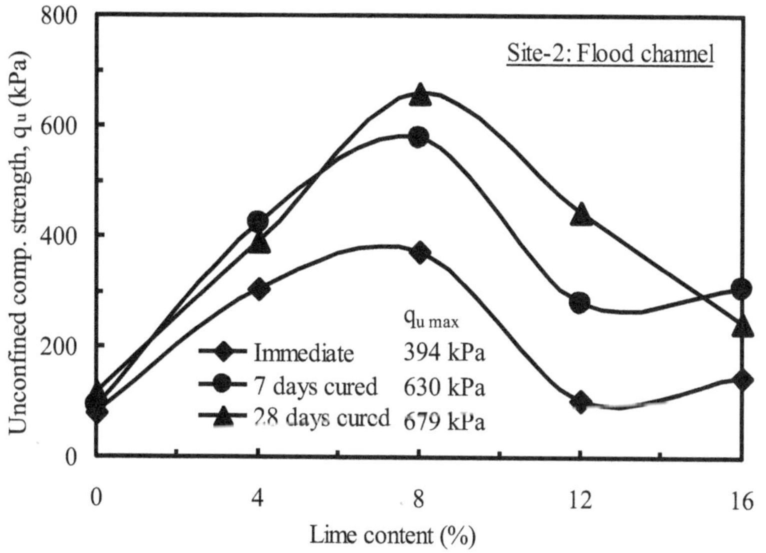

Fig. 9. Strength variation of lime stabilized dredged soil with change in lime content

during first 7 days and then increases slowly almost at a constant rate [34]. With increase in curing period, the strength of the treated soil increases however, there is a rapid rate of gain of strength for initial 7 days of curing period, (394 to 630 kPa) due to the fact that all pozzolanic reactions get completed within this period. After 7 days there is a marginal increase in strength of the soil from 630 to 679 kPa.

Thompson [64] stated that addition of lime to fine-grained soils yields a substantial increase in cohesion and minor improvement in internal friction angle. However, some researchers have shown significant increase in internal friction angle (ϕ) as well up to a certain optimum value. Beyond the optimum value, addition of lime has no effect on internal friction angle [65]. Direct shear tests were conducted on untreated and treated test specimens. The test results reveal that in-situ dredged soil exhibits soft consistency. The angle of internal friction of 16° indicates loose denseness of the soil. Mohr Coloumb failure envelopes (Fig. 10) show that lime treatment drastically improves both cohesion (c) and internal friction angle (ϕ) due to pozzolanic reactions and cation exchange between lime and soil particles. The soil state changes from loose to dense state. The shear parameters showed decreasing trend beyond 8% lime content. The variation of c and ϕ with increase in lime content is shown in Fig. 11.

The increase in cohesion can be expressed as a function of unconfined compressive strength directly, as given by Thompson [64]

$$c = 9.3 + 0.292c \tag{1}$$

where,

c is cohesion (psi), and
σ_c is unconfined compressive strength (psi)

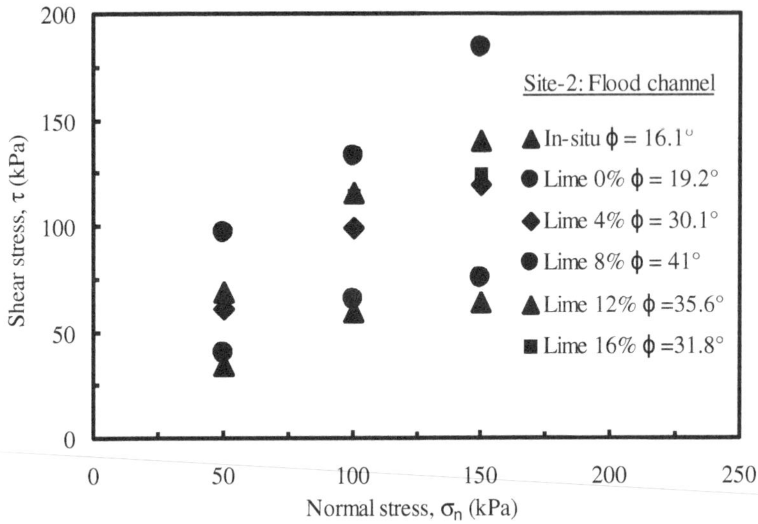

Fig. 10. Mohr Coloumb Failure envelope of lime stabilized dredged soil

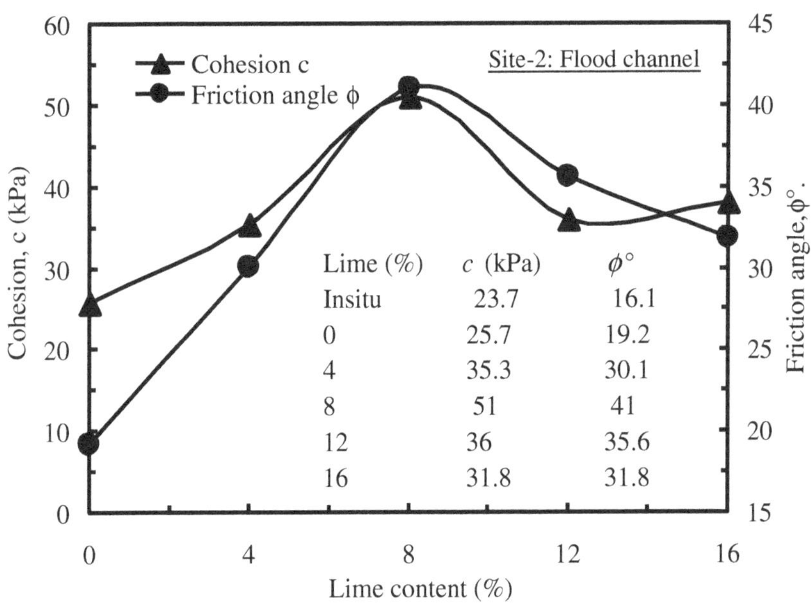

Fig. 11. Variation of c and ϕ with change in lime content

5 Micro-structural Behavior Using Scanning Electron Microscope (SEM)

A scanning electron microscope (SEM) gives us a specific idea about the shape of clay particles by producing surface images of a soil specimen at a microscopic level [66]. Various researchers have studied the micro-structure of clay particles by using scanning electron microscopy technique [67–69]. Keshawarz [70] reported that the particles of natural soil appear as a blocky arrangement of loosely packed particles. Ismail [71] stated that the natural untreated expansive clayey silt showed flaky arrangements of clay particles as a matrix between the silt and fine sand grains. Ali [72] said that clay minerals in the natural soil exist in the form of large or small aggregates or clusters whose size depends on soil composition and physical state. Colins and McGown [73] distinguished micro-fabric features of natural soil into three principal forms:

1. Elementary particle arrangements: This constitutes single forms of particle interaction at the level of individual clay, silt, or sand particles or interaction between small groups of clay platelets or clothed silt and sand particles.
2. Particle assemblages: These are units of particle organization having definable physical boundaries and specific mechanical function. They consist of one or more forms of elementary particle arrangements or smaller particle assemblages. Particle assemblages include matrices, aggregations and connectors. Al-Rawas [74] said clay matrices may be honeycombed with interference from embedded silt or sand grains. Aggregations may be in the form of disturbed parallel clay arrangements, with embedded silt or sand. Connectors are assemblages which link silt or sand grains and in certain cases even aggregations.
3. Pores: They are spaces formed within and between elementary particle arrangements and particle assemblages. On the basis of cluster theory pores have been classified as intra aggregate pores and inter aggregate pores [75, 76].

Fig. 12. SEM image of in-situ soil

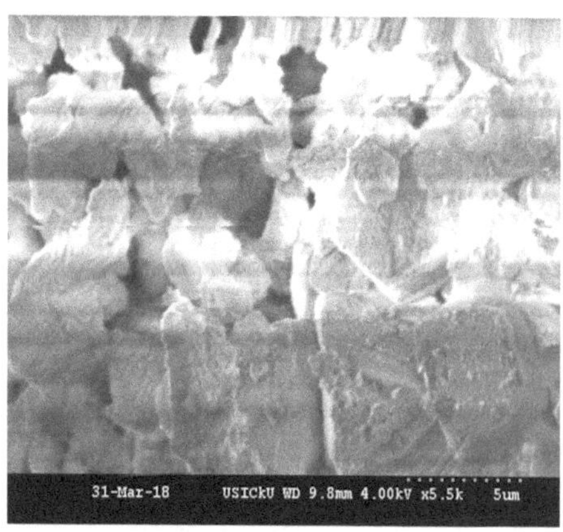

Fig. 13. SEM image of hydrated lime

However, other researchers like Nagaraj [75], Aylmore and Quirk [77] and Olsen [78] have reported that the basic element of the microstructure of natural clay is not the single platelet, but domains, composed of various aggregated platelets.

In the present study, samples for SEM analysis were taken from the shear failure plane of tested UCS samples in the form of pat of size 10 mm × 10 mm × 7.5 mm. Small specimens of both untreated and treated clay samples were prepared and the fractured surfaces of the air dried specimens were coated with gold before scanning. The SEM images of lime treated samples were obtained to see the effect of lime on microstructure of the clay particles. The SEM testing of the natural sample of dredged material revealed presence of loose flaky particles in deflocculated form with distinct open spaces having weak inter-particle bonds, forming a dispersed structure. Such dispersed and deflocculated structure imparts weak strength to the soil mass (Fig. 12) [79].

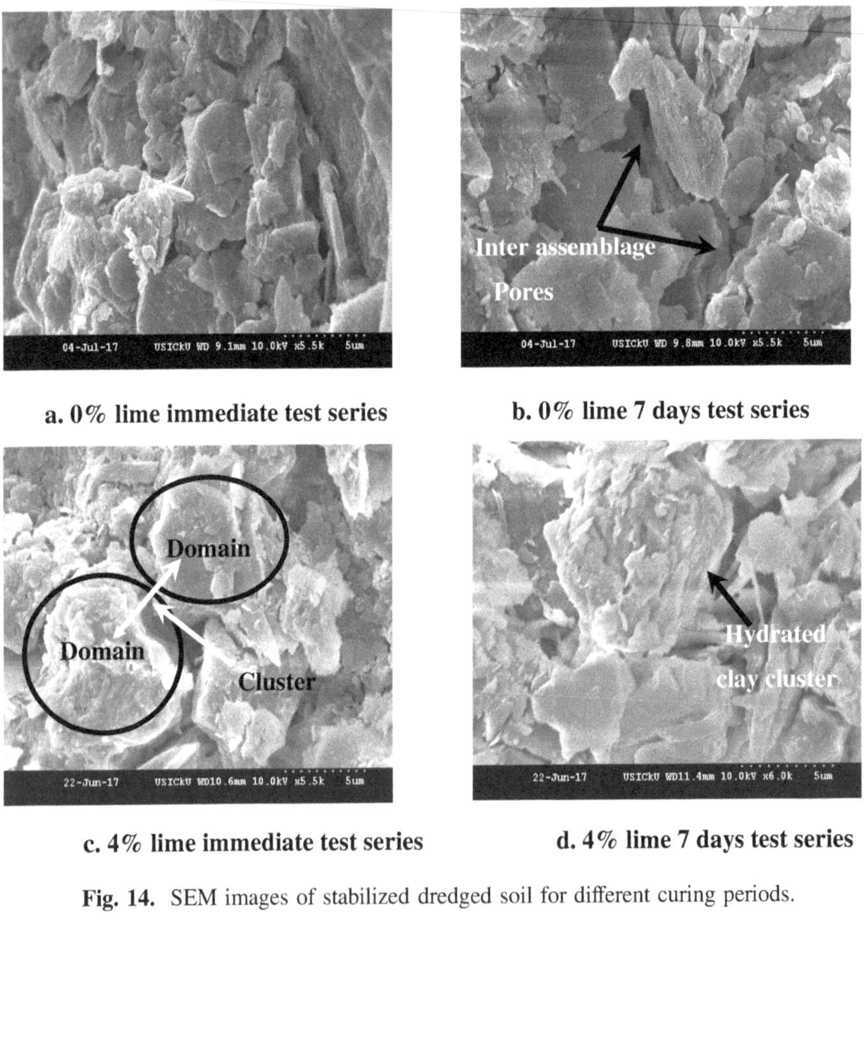

a. 0% lime immediate test series **b. 0% lime 7 days test series**

c. 4% lime immediate test series **d. 4% lime 7 days test series**

Fig. 14. SEM images of stabilized dredged soil for different curing periods.

e. 8% lime immediate test series

f. 8% lime 7 days test series

g. 12% lime immediate test series

h. 12% lime 7 days test series

i. 16% lime immediate test series

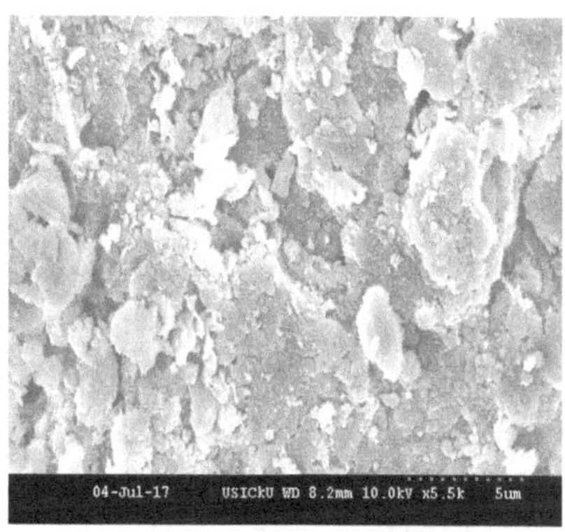

j. 16% lime 7 days test series

Fig. 14. (*continued*)

With the addition of lime, several clay particles stack over one another forming coherent masses (Fig. 13). They exhibit face to face association leading to the formation of well aggregated clay particles in easy to spot masses of lime clay clusters (Fig. 14c). The addition of lime depicts the presence of cementing products in the pore system with a well knit network between the clusters. Kawamura and Diamond [80] have reported formation of cementing products ($CaO–SiO_2–H_2O$, $CaO–Al_2O_3–H_2O$) in lime stabilized clays. The cementing compounds bond clusters together forming a bridge between clusters called as bound bridge [81].

The pozzolanic action of lime increases with increase in curing period up to 7 days which gives rise to the formation of cementitious products like calcium silicate hydrate (C.S.H), calcium aluminate hydrate, (C.A.H) and calcium alumino silicate hydrate (C.A.S.H). These compounds fill the pores and enhance inter particle bonding, thereby increasing the strength [82]. The lime clay clusters tend to be larger due to abundance of these compounds with time (Fig. 14d). The hydration products coat the soil particles and join them together making the voids less distinct (Fig. 14f). Further, with increase in lime content the inter particle bonding shows tangible gain. However, beyond the optimum value of lime content, the SEM micrographs show definite disunion among the clay particles due to the presence of repulsive forces (Fig. 14g and i).

6 Conclusions

The key findings of the present study can be summarized as under:

1. The addition of lime to the dredged soil significantly improved the mechanical properties of soil.
2. The plasticity index decreased up to 12% lime content, after which it increased making the soil more workable up to this value.
3. The lime treatment flattens the compaction curve thereby ensuring that a given percentage of the prescribed density can be achieved over a wider range of moisture contents so that relaxed moisture control specifications are possible.
4. The unconfined compressive strength increased by about 6 times than the untreated soil. The optimum value of lime content was observed at 8% beyond which the compressive strength decreased.
5. The Mohr–Coloumb parameters showed significant gain, improving the consistency from soft to hard state and soil state from very loose to dense. However, the addition of lime changed the failure pattern of dredged soil from ductile to brittle.
6. The curing period strengthens the inter particle bonds due to pozzolanic reactions. With increase in the percentage of lime, the flocculation increases with reduction in void spaces forming large particles.
7. The SEM analysis shows that the mixing of lime changes the structural arrangement of clay particles from deflocculated form to flocculated from. There is the development of cementitious products that increases the inter particle bonding.

Acknowledgments. The investigation reported in this paper forms a part of the Research Project work on reinforced soils at NIT Srinagar. Thanks are due to the faculty of Geotechnical Engineering division and supporting staff of the Soil Mechanics laboratory for their timely help during the course of this investigation.

References

1. Mir, B.A.: Some studies on geotechnical characterization of dredged soil for sustainable development of Dal Lake and environmental restoration. Int. J. Tech. Res. Appl. **12**, 04–09 (2015)
2. Mir, B.A.: Need for dredging and dredged material characterization from Dal Lake Srinagar - an overview. In: Proceedings of IGC-2013, vol. 1, Th-1: Paper No. 39, pp. 1–9, IIT Roorkee, India (2005)
3. Ghobadi, M.H., Abdilor, Y., Babazadeh, R.: Stabilization of clay soils using lime and effect of pH variations on shear strength parameter. Bull. Eng. Geol. Environ. **73**, 611–619 (2013)
4. Virgina Govt. Management of dredged material. Solid waste interpretive guidance statement
5. Jones, R.A., Lee, G.F.: The significance of dredging and dredged material disposal as a source of nitrogen and phosphorus for estuarine waters. In: Estuaries and Nutrients, pp. 517–530. Humana Press, Clifton (1981)
6. Maher, A.D., Douglas, W.S., Jafari, F., Pecchioli, J.: The processing and beneficial use of fine grained dredged material, a manual for engineers, New Jersey, USA (2013)
7. Prakash, A., Paul, A.: A study on stabilization of marine dredged soil using quarry dust. Int. J. Sci. Eng. Res. **4**(3), 83–86 (2016)
8. Roger, T.: Dredged material as a natural resource concepts for land improvement and reclamation. Saucier Environmental Effects Laboratory U.S. Army Engineer Waterways Experiment Station (1976)
9. Millrath, K., Kozlova, S., Shimanovich, S., Meyer, C.: Beneficial Use of Dredge Material. Department of Civil Engineering and Engineering Mechanics, Columbia University, New York (2001)
10. Stephens, S.R., Alloway, B.J., Carter, J.E., Parker, A.: Towards the characterisation of heavy metals in dredged canal sediments and an appreciation of availability two examples from the UK. Environ. Pollut. **113**, 395–401 (2001)
11. Peng, J.F., Song, Y.H., Yuan, P., Cui, X.Y., Qiu, G.L.: The remediation of heavy metals contaminated sediment. J. Hazard. Mater. **161**, 633–640 (2009)
12. Fu, J., Zhao, C., Luo, Y., Liu, C., Kyzas, G.Z., Luo, Y., Zhao, D., An, S., Zhu, H.: Heavy metals in surface sediments of the Jialu River, China: their relations to environmental factors. J. Hazard. Mater. **270**, 102–109 (2014)
13. Yu, K.C., Tsai, L.J., Chen, S.H., Ho, S.T.: Chemical binding of heavy metals in anoxic river sediments. Water Res. **35**, 4086–4094 (2001)
14. Pang, H.J., Lou, Z.H., Jin, A.M., Yan, K.K., Jiang, Y., Yang, X.H., Arthur Chen, C.T., Chen, X.G.: Contamination, distribution, and sources of heavy metals in the sediments of Andong tidal flat, Hangzhou bay. China Cont. Shelf Res. **110**, 72–84 (2015)
15. Bastami, K.D., Bagheri, H., Kheirabadi, V.: Distribution and ecological risk assessment of heavy metals in surface sediments along southeast coast of the Caspian Sea. Mar. Pollut. Bull. **81**, 262–267 (2014)
16. Guan, Q., Wang, L., Pan, B., Guan, W., Sun, X., Cai, A.: Distribution features and controls of heavy metals in surface sediments from the riverbed of the Ningxia-Inner Mongolian reaches, Yellow River, China. Chemosphere **144**, 29–42 (2016)

17. Li, X., Cao, Y., Qi, L., Shu, F.: The distribution characteristics of heavy metals in Guiyang urban soils. Chin. J. Geochem. **31**, 174–180 (2012)
18. DOER. Dredged material characterization tests for beneficial use suitability, Technical Note DOER-C2 (1999)
19. Cappuynsa, V., Swennen, R.: Oxidation of Anoxic Soils and Sediments How Can We Evaluate the Risk of Heavy Metal Release. Nova Science Publishers, Belgium (2009)
20. Kelly, W.R., Wehrmann, H.A., Holm, T.R., Talbott, J.L., Skowron, L.M.: Subsurface movement of zinc from contaminated dredge spoils at a periodically flooded site. Environ. Geol. **45**, 23–34 (2003)
21. Mir, B.A.: Some studies on the physical and mechanical behavior of dredged soil from Jhelum river, flood spill channel of river Jhelum, Srinagar. Acta Ing. Civ. **01**(01), 1–7 (2016)
22. Beeghly, H.J., Shrock, M.: Dredge material stabilization using the pozzolanic or sulfo pozzolanic reaction of lime byproducts to make an engineered structural fill. In: World Coal Ash Conference, Lexington, USA (2009)
23. Bartos Jr., M.J.: Classification and engineering properties of dredged soil. Technical report D-77-18, U.S. Army Waterways Experiment Station, Vicksburg (1977)
24. Mir, B.A., Mir, F.A.: Applications and geotechnical evaluation of dredged soil obtained from Dal Lake in Srinagar. In: Proceedings of National Conference on Soils and Their Applications in Civil Engineering, pp. 26–37 (2004)
25. Aarninkhof, S., Luijendijk, A.: Safe disposal of dredged material in an environmentally sensitive environment. Port. Technol. Int. **47**, 39–45 (2010)
26. Haliburton, T.A.: Guidelines for dewatering/densifying confined dredged material, synthesis report. Technical Report DS-78-11, US Army Engineer Waterways Experiment Station, Vicksburg, Mississippi (1978)
27. U.S. Environmental Protection Agency and U.S. Army Corps of Engineers. Identifying, planning, and financing beneficial use projects using dredged material, EPA842-B-07-001 (2007)
28. Al-Rawas, A.A., Hago, A.W., Al-Sarmi, H.: Effect of lime, cement and sarooj (artificial pozzolan) on the swelling potential of an expansive soil from Oman. Build. Environ. **40**(5), 681–687 (2005)
29. Mcdowell, C.: Stabilization of soils with lime, lime fly ash and other reactive materials. Highw. Res. Road Bull. **231**, 60–66 (1959)
30. Sherwood, P.T.: Soil Stabilization of Cement and Lime, State of art review, Transport Research Laborotary. Her Majesty's Stationary Office, London (1993)
31. Perry, J., Snowden, R.A., Wilson, P.E.: Site investigation for lime stabilization of highway works. Transport Research Laboratory, Department of Transport, Berkshire RG40 3GA United Kingdom, pp. 85–96 (1995)
32. Anon: Lime Stabilization Construction Manual, 8th edn. National Lime Association, Arlington (1985)
33. Anon: Lime Stabilization Manual. British Aggregate Construction Materials Industry, London (1990)
34. Bell, F.G.: Lime stabilization of clay minerals and soils. Eng. Geol. **42**(4), 223–227 (1996)
35. Ismail, H.A.H.: Treatment and improvement of geotechnical properties of different soft fine grained soils using chemical stabilization. Martin Luther University of Halle- Wittenberg (2006)
36. Holt, C.C., Freer-Hewish, R.J.: Lime treatment of capping layers in accordance with the current specification for highway works. In: Proceedings of Seminar on Lime Stabilization. Loughborough University, Civil and Building Engineering Department, London, UK, pp. 51–61 (1996)

37. Muhuntham, B., Sarioserri, F.: Interpretation of geotechnical properties of cement treated soils, Washington State University, Department of Civil and Environmental Engineering (2008)
38. Croft, J.B.: The influence of soil mineralogical composition on cement stabilization. Geotechnique **17**, 119–135 (1967)
39. Rafalko, S.: A thesis work on rapid soil stabilization of soft clay soils for contingency airfields. Faculty of Virgina Polytechnic Institute and State University (2006)
40. Basma, A.A., Tuncer, E.R.: Effect of lime on volume change and compressibility of expansive clays. Transportation Research Record, Washington, D.C., TRR No. 1296:54–61 (1991)
41. Eades, J.L., Grim, R.E.: A quick test to determine the lime requirements of lime stabilization. Highw. Res. Rec. **139**, 61–72 (1966)
42. Little, D.N.: Handbook for Stabilization of Pavement Subgrades and Base Courses with Lime. Kendall/Hunt Publishing Company, Dubuque (1995)
43. Locat, J., Berube, M.A., Choquette, M.: Laboratory investigation on the lime stabilization of sensitive clay, shear strength development. Can. Geotech. J. **27**(3), 77–91 (1990)
44. UK Marine Special Areas of Conservation Project. Environmental impacts of maintenance dredging and disposal (2001)
45. Wei, Z.: Environmental impact of dredging and IWT development (Chap. 3). College of Harbour, Waterway and Coastal Engineering, Hohai University, Nanjing, China
46. Muntohar, A.S., Hantoro, G.: Influence of Rice Husk Ash and Lime on engineering properties of a clayey sub-grade, ejge paper 2000-094 (2000)
47. IS: 1498. Classification and identification of soils for general engineering purposes. Bureau of Indian Standards, New Delhi (1970)
48. IS: 2720-part 1. Indian Standard Code for preparation of soil samples. Bureau of Indian Standards, New Delhi (1980)
49. IS: 2720-part 5. Method of test for soils: determination of atterberg limits. Bureau of Indian Standards, New Delhi (1985)
50. IS:2720-part 10. Method of test for soils: determination of shear strength parameter by unconfined compression test. Bureau of Indian Standards, New Delhi (1973)
51. IS: 2720-part 39. Method of test for soils: determination of shear strength parameter by direct shear test. Bureau of Indian Standards, New Delhi (1977)
52. Mir, B.A.: Some studies on Engineering behavior of cement stabilized dredged soil from flood spill channel of Jhelum river, Srinagar, 50th IGC Pune, India (2017)
53. Elkady, T.A.: The effect of curing conditions on the unconfined compressive strength of lime treated expansive soils. J. Road Mater. Pavement Des. **17**(1), 52–69 (2016)
54. Uddin, K., Balasubramaniam, A.S., Bergardo, D.T.: Engineering behavior of cement treated Bangkok soft clay. Geotech. Eng. J. **28**(1), 89–119 (1997)
55. Bushra, I., Robinson, R.G.: Strength behavior of cement stabilized marine clay cured under stress, Geotrendz, IGC 2010 (2010)
56. Mir, B.A., Amin, F., Majid, B.: Some studies on physical and mechanical behavior of dredged soil from flood spill channel of Jhelum river, Srinagar. Acta Ing. Civ. **01**(01), 01–07 (2016)
57. IS: 2720-part 7. Method of test for soils: determination of water content-dry density relation using light compaction. Bureau of Indian Standards, New Delhi (1980)
58. Skels, P., Nielsen, T.I., Bondars, K.: Quick lime stabilization of fine grained marine sediments in low temperature areas. In: 3rd International Conference on Civil Engineering Proceedings I Building Materials, pp. 71–77 (2011)

59. Mir, B.A.: Some studies on the effect of fly ash and lime on physical and mechanical properties of expansive clay. Int. J. Civil Eng.-Trans. B: Geotech. Eng. Geotech. **13**(3&4B), 203–212 (2015)
60. Jan, O.Q., Mir, B.A.: Mechanical behavior of cement stabilized dredged soil. Glob. J. **14**(4) (2017)
61. Nicholson, P., Kashyap, V., Fuji, C.: Lime and fly ash admixture improvement of tropical Hawaiian soils. Transportation Research Record, Washington, D.C., No. 1440, pp. 71–78 (1994)
62. Hussain, M., Dash, S.K.: Influence of Lime on the Compaction Behavior of Soils ICE Publishing Environmental Geotechnics (2015)
63. Mitchell, J.K., Hooper, D.R.: Influence of time between mixing and compaction on properties of lime stabilized expansive clay. High. Res. Board Bull. **304**, 14–31 (1961)
64. Thompson, M.R.: Shear strength and elastic properties of lime-soil mixtures. Highway Research Record, Washington, D.C., vol. 139, pp. 1–14 (1966)
65. Szendefy, J.: Impact of the soil-stabilization with lime. In: Proceedings of the 18th International Conference on Soil Mechanics and Geotechnical Engineering, Paris (2013)
66. Wischnitzer, S.: Introduction to Electron Microscopy, 2nd edn. Pergamon, Print, New York (1970)
67. Gillott, J.E.: Study of the fabric of fine-grained sediments with the scanning electron microscope. J. Sediment. Petrol. **39**, 90–105 (1969)
68. McKyes, E., Yong, R.N.: Three techniques for fabric viewing as applied to shear distortion of clay. Clays Clay Miner. **19**(5), 289–293 (1971)
69. Abduljauwad, S.N.: Improvement of plasticity and swell potential of calcareous expansive clays. Geotechn. Eng. J. **26**(1), 3–16 (1995)
70. Keshawarz, M.S., Dutta, U.: Stabilization of south texas soils with fly ash. In: Sharp, K.D. (ed.) Proceedings of the Geotechnical Special Publication No. 36, ASCE, Dallas, Texas, United States, pp. 30–42 (1993)
71. Ismail, H.A.H.: Cement kiln dust chemical stabilization of expansive soil exposed at El-Kawther, Quarter, Suhag region, Egypt. Int J. Geosci. (2013)
72. Ali, H.M., Liu, H., Gao, Y., Peng, J., Li, T.: Microstructure study of soft clayey soil under consolidation by vacuum preloading method. Geomech. Geoeng. **2**(2), 97–108 (2007)
73. Collins, K., McGown, A.: The form and function of microfabric features in a variety of natural soils. Geotechnique **24**, 233–254 (1974)
74. Al-Rawas, A.A.: Microfabric and mineralogical studies on the stabilization of an expansive soil using cement by-pass dust and some types of slag. Can. Geotech. J. **39**, 1150–1167 (2002)
75. Nagaraj, T.S., Vatasala, A., Srinivasa, M.B.R.: Discussion on change in pore size distribution due to consolidation of clays by F.J. Griffith and R.C. Joshi. Geotechnique **40**(2), 303–305 (1990)
76. Mitchell, J.K.: Fundamentals of Soil Behavior, p. 422. Wiley, New York (1976)
77. Aylmore, L.A.G., Quirk, J.P.: Domain or turbostratic structure of clays. Nature **187**, 1046 (1960)
78. Olsen, H.W.: Hydraulic flow through saturated clays. Clay Clay Miner. **9**(2), 131–161 (1962)
79. Sides, G., Barden, L.: The microstructure of dispersed and flocculated samples of kaolinite, illite, and montmorillonite. Can. Geotech. J. **8**(3), 391–399 (1971). NRC Research Press, Canadian Science Publishing

80. Kawamura, M., Diamond, S.: Stabilization of clay soils against erosion loss. Clay Clay Miner. **23**, 443–451 (1975)
81. Katsutada, O., Chirdchanin, M., Masakazu, K.: Investigation on Microstructure and Strength of Lime and Cement Stabilized Ariake Clay (2001)
82. Horpibulsuk, S., Rachan, R., Chinkulkijniwat, A., Raksachon, Y., Suddeepong, A.: Analysis of strength development in cement-stabilized silty clay from microstructural considerations. Constr. Build. Mater. **24**, 2011–2021 (2010)

Finding Engineering Solutions in Collapsible Soils, Before and After Construction

Peter Scott(✉)

BuroHappold Engineering, Bath, UK
peter.scott@burohappold.com

Abstract. Two projects undertaken in Saudi Arabia are contrasted and issues that were identified with the founding strata are discussed. In both cases the main condition in the soil was its susceptibility to collapse, specifically under inundation. Interestingly although one deposit was within wind-blown deposits, a form of deposition that is known to potentially give rise to collapsible soils, the other project was on a site underlain by wadi deposits.

The first case study looks at an existing building and how signs of structural distress became evident over a number of years. A forensic investigation was undertaken to determine the causes of this problem. The second case study is focused on a prestigious museum development in Riyadh. Initial investigations suggested that the building could be founded on shallow strip footings, in view of the apparent competency of the underlying wadi deposits. Subsequent in-situ testing identified soil collapse within these strata.

Concluding remarks are provided on investigation techniques to be used to identify these types of problem soils and the value of engineers using observational skills, rather than focusing on the relatively indiscriminate use of routine field and laboratory testing.

1 Introduction

Over the last three decades the author has been involved in a number of projects across the Middle East. Prior to, and during this period, the construction industry has generally focused on aggressive ground and groundwater conditions. Although academic papers have identified the occurrence and mechanisms that give rise to collapsible soils in the region little has been written on projects where the phenomena has been identified and overcome. This paper seeks to partly rectify this by reference to two projects in Saudi Arabia. Although for most people collapsible soils are synonymous with wind-blown deposits for one of these projects no significant wind-blown material was identified.

For both projects a forensic investigation had to be undertaken, as in neither case was there obvious evidence that collapsible soils were to prove to be a problem. It was however possible using good observational skills, along with relatively simple in-situ test procedures, to identify the soils liable to collapse. Once the source of potential collapse was found differing approaches had to be used to address matters, particularly as at one of the sites the building had been in existence for some years before the problem was uncovered.

H. Shehata and H. Poulos (Eds.): GeoMEast 2018, SUCI, pp. 154–162, 2019.
https://doi.org/10.1007/978-3-030-01917-4_12

2 Saudi National Museum, Riyadh

The Saudi National Museum, forming part of the King Abdul Aziz Historical Centre, was constructed to coincide with the 100th anniversary of the founding of the modern state of Saudi Arabia. The project incorporated the historically significant Murabba Palace and the adjacent Darat Al Malik Abd Al Aziz (Fig. 1).

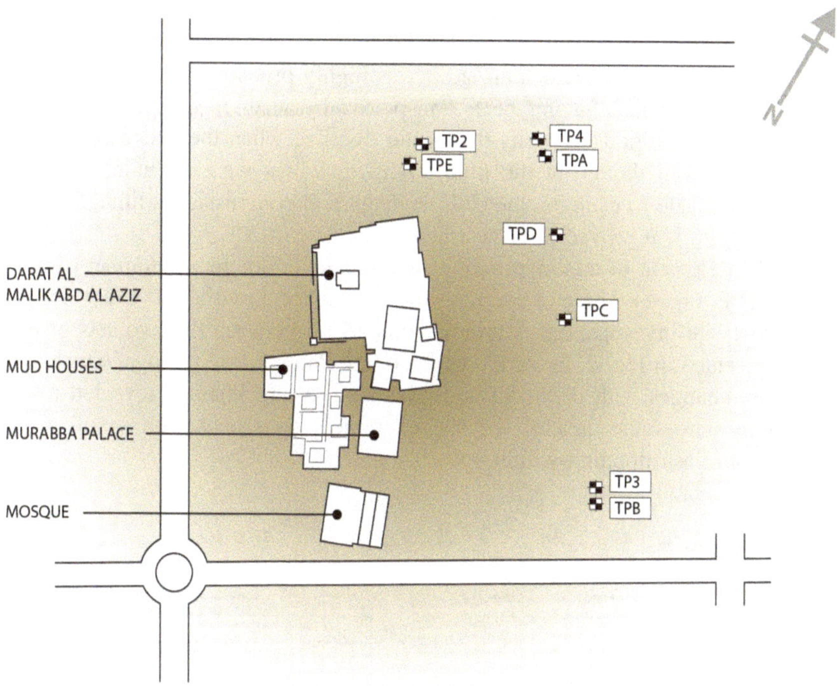

Fig. 1. Saudi National Museum (SNM) site layout and selected trial pit locations

2.1 Phase 1 Site Investigation

A site investigation was undertaken by Dames and Moore in 1996. This broadly confirmed an earlier investigation by Al Muhandis in 1986, with around 7 to 13 m of superficial deposits overlying limestone of the Arab Formation. The borehole logs indicated soils were predominantly sands and gravels, with occasional cohesive inclusions. Fill, up to 1.5 m thick, was found to overlie the natural deposits towards the northern end of the site.

During the laboratory testing phase of this work the possibility that the soils may be collapsible was raised by the contractor. To try and prove, or otherwise, the possibility of this two strategies were considered. Firstly to determine if cementitious deposits, such as carbonates, were present in the soils; they could be liable to dissolution and subsequently lead to soil collapse. All the samples tested however gave carbonate contents of close to zero. Secondly consideration was given to using oedometer tests to quantify soil collapse. The sandy nature of the deposits led the author to conclude that it would be extremely unlikely that undisturbed samples could sensibly be obtained and so this element of work was not pursued.

2.2 Phase 2 Site Investigation

The previous investigation work had seemingly shown that the natural deposits were compact and could comfortably carry a design load of 250 kPa. Although the boreholes had not proven the soils were stratified it was highly probable that the soils had been deposited under water, having been transported by wadi flow. There was also no evidence of cementitious deposits that could dissolve when the site was irrigated. In spite of this, even though at the time tender drawings were about to be issued, a decision was made to examine the soils in detail and commission a limited additional testing program if it proved to be warranted.

The first element of this investigation work entailed logging a number of trial pits (TPA to TPE. Fig. 1). Three of the pits were dug adjacent to trial pits excavated during the Phase 1 site investigation. A typical comparison between the two sets of trial pit logs is presented in Fig. 2. In stark contrast to the Phase 1 trial pit logs, which showed no distinct changes with depth, those undertaken during Phase 2 revealed a layered stratigraphy across the area of the site examined, consistent with a soil sequence derived from flash flooding within a wadi.

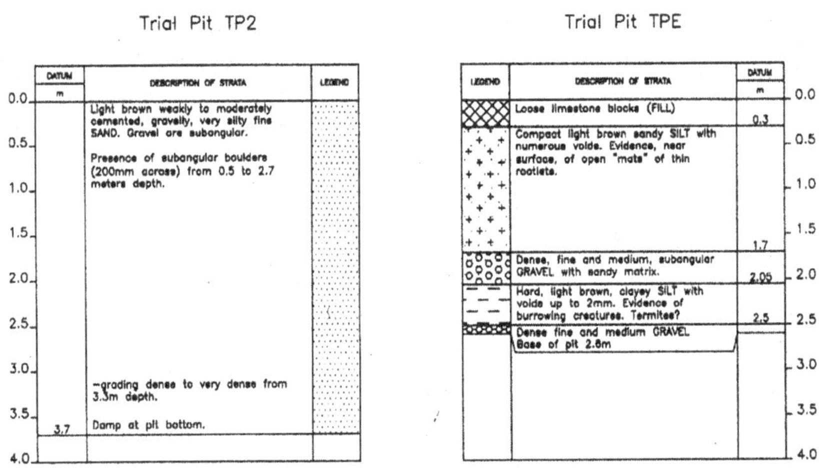

Fig. 2. Comparison of trial pit logs at SNM

Close examination, particularly of the sandy and clayey silt deposits in the trial pits (see Fig. 2), revealed that these soils were riddled with numerous pin prick sized voids, the number of which appeared to diminish with depth. The evidence of how these voids were formed was specifically found in trial pit TPE, where root systems from bushes around the trial pit were revealed as myriads of fine roots extending downwards into the soil. A secondary series of larger voids, up to 2 mm in diameter, were also found. These probably arose from termites, or other burrowing creatures, which were observed at about 2 m below ground in trial pit TPE.

With clear evidence of voids being present in the superficial deposits a series of plate load tests were undertaken, to examine whether the voids would collapse under foundations when exposed to water. Three horizons were specifically examined:

1. An upper sandy silt horizon - Trial pit TPB at 1 m below ground level.
2. A lower sandy silt horizon - Trial pit TPE at 2.2 m below ground level
3. A gravel horizon, separating the upper and lower silt horizons - Trial Pit TPC at 1.2 m below ground level.

2.3 Plate Load Tests

With the intention to irrigate the landscaped grounds within the project area, following completion of the works, it was expected that the majority of the sedimentary deposits would become saturated. To take account of this anticipated long term change the plate load tests generally followed the sequence given below:

- The 600 mm diameter plate was loaded, in four equal stages, up to an applied bearing pressure of about 290 kPa. Load was maintained, at each stage, until the rate of settlement did not exceed 0.25 mm/hr
- With the pressure of 290 kPa still applied the test pit was inundated and settlement monitored for a minimum additional period of 48 h.
- Once the test was complete, the soils were excavated from beneath the plate test area and moisture content samples were taken at 0.25 m centres.

Results obtained from plate tests, on the three selected horizons, are presented in Fig. 3.

All the plate tests carried out on the dry soils gave very low settlements, not exceeding 3 mm. When the soils were wetted up by inundation of the plate test area, the movements increased dramatically. It became evident that the settlements measured were well in excess of that which would be acceptable in the final structure. For the test on the gravel, when the soils were excavated beneath the plate, silt deposits were found only 500 mm down. It was concluded that the settlement obtained was primarily due to the performance of the silt, rather than the gravel.

2.4 Foundation Design

It was apparent from the plate test results that ancient root systems had developed a soil structure that was susceptible to collapse when saturated. Unfortunately it had not been possible to prove the founding level below which collapse would not occur; oedometer tests on samples at depth were unsuccessful as undisturbed samples could not be obtained.

To assist in determining an appropriate founding level consideration was given to existing and previous groundwater levels. The reasoning being that roots would not

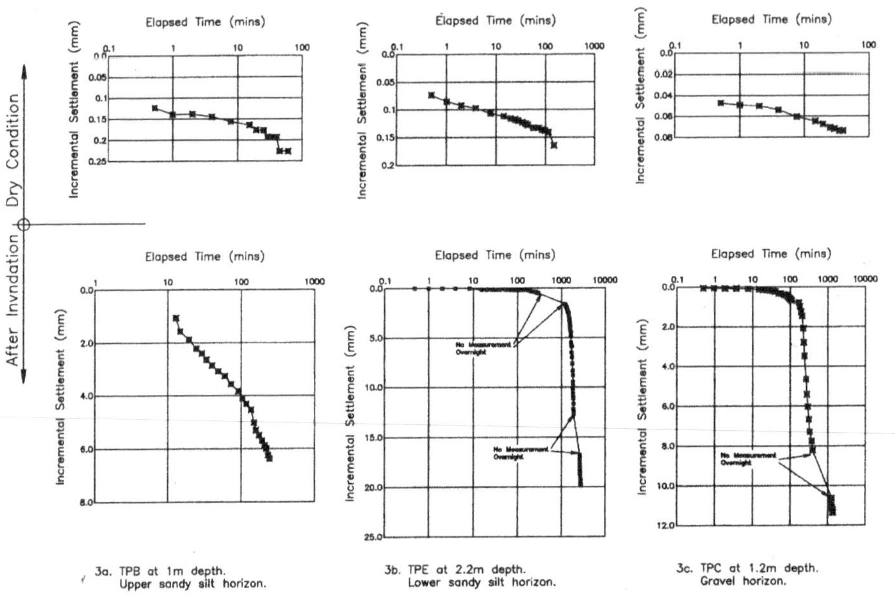

Fig. 3. Plate tests before and after inundation at SNM

have penetrated below the water table. At the time of the investigation the water table was 5 to 6 m below ground level (bgl). This was found to be below the historic groundwater level of 3 mbgl, as pumping was being carried out to lower the water level at the underpasses of the nearby King Fahd highway.

From the foregoing a foundation strategy was evolved:

1. Foundations were to be designed to impose a maximum load of 250 kPa on the founding strata.
2. The founding level was set at a minimum level of 3.0 m below existing ground level; the historic groundwater level
3. Plate tests, replicating the procedure used in Phase 2 of the site investigation, would be undertaken at the proposed founding levels to prove that collapse settlement would not occur.
4. If collapse settlement happened during any of the plate load tests, then foundations should be taken down another 0.5 m and the test procedure repeated. Alternatively, if gravels were found at a suitable depth footings could be placed at this level, once the deposit was proven to be of an adequate thickness.

As the sequence of silts, sands and gravels could give rise to perched water tables across the site the footings were also designed to act as water cut-offs around the perimeter of the main museum building. By this process water could be prevented from encroaching too close to the underside of the building floors and ground bearing slabs could be used.

Subsequent plate load tests during the foundation works the following year confirmed the strategy was successful, with further tests being only minimally affected by inundation.

3 Signal School, Taif

Taif is situated in the hills about 80 km east of Jeddah. The site of the Signal School is surrounded by rocky outcrops but the headquarters building itself is on level ground. Around the building, apart from the side adjacent to a parade ground, vegetation and trees have been planted. On the whole the vegetation was mature and at the time of the investigation was generally flourishing. Prior to the investigations undertaken by the author the headquarters building had been experiencing several years of evident differential movement between various sections of the structure (Fig. 4).

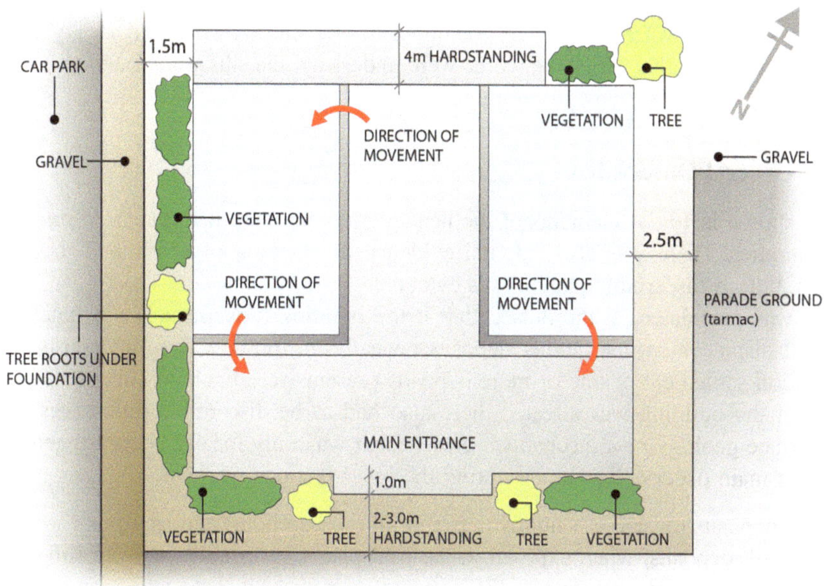

Fig. 4. Building layout and damage at Taif

3.1 Ground Conditions

From undertaking three short boreholes and two trial pits on site the ground conditions were found to consist of the following superficial deposits

- Fill - Compact sand and gravel to about 1.0 to 1.5 m below ground level
- Wadi deposits - Dense brown sands and gravels
- Wadi/Wind blown deposits - Very stiff light brown silt

The actual stratification of these materials is detailed in a summary trial pit log given in Table 1. Boreholes showed the materials were generally consistently very dense and sandy from about 4.0 m below ground level.

Table 1. Summary Trial Pit Log at Taif

0–0.6 m	Damp dark brown sandy SILT and angular to sub angular GRAVEL (Fill)
0.6–0.9 m	As above, but dry and brown (Fill)
0.9–1.5 m	Dense, brown, angular to sub-angular fine medium and coarse GRAVEL with some cobbles in a sandy silt matrix
1.5–1.8 m	Dense, grey-brown, silty SAND and much gravel
1.8–3.9 m	Stiff brown slightly sandy SILT
3.9–4.0 m	Dense, orange brown, silty SAND with some gravel

No standing water levels were identified during the investigation. Heavy rainfall did occur most nights whilst the works were underway; the silts were found to be damp when excavated the following morning.

3.2 General Observations

As is shown in Fig. 4 elements of the headquarters building had displaced relative to one another. There was also external evidence of cracking of lintels and roof overhangs, but no discernible pattern was evident.

It was postulated, at the outset, that if the building movement was a function of ground displacement then either deposits beneath the footings were undergoing consolidation settlement or one or more deposits present were liable to collapse. As only parts of the building was affected there also had to be discernible differences in the subsurface geology or a mechanism present that was only locally being triggered.

The main observations made during the investigations were:

- Silt deposits existed at shallow depth in all the boreholes and trial pits
- The silt deposits, when exposed in the trial pits, were usually dry and stiff
- The standard penetration test (SPT) samples in the silt were moist, and the SPT values were lower than would have been expected from the trial pit inspections.
- When a trial pit was excavated following overnight rain the underlying silt deposits were found to be damp.

The evidence was indicating the silts were prone to rapid moisture changes. There was also some indication that these deposits exhibited a loss of strength when they became moist.

3.3 Plate Tests

A similar form of plate load test was undertaken as Taif as had been done at the Saudi National Museum. In this instance the foundation pressure was estimated to be about 180 kPa. Figure 5 shows the results of two tests which were both inundated. In one test only half the load of the other test was used to check sensitivity of the soil to changes in applied pressure.

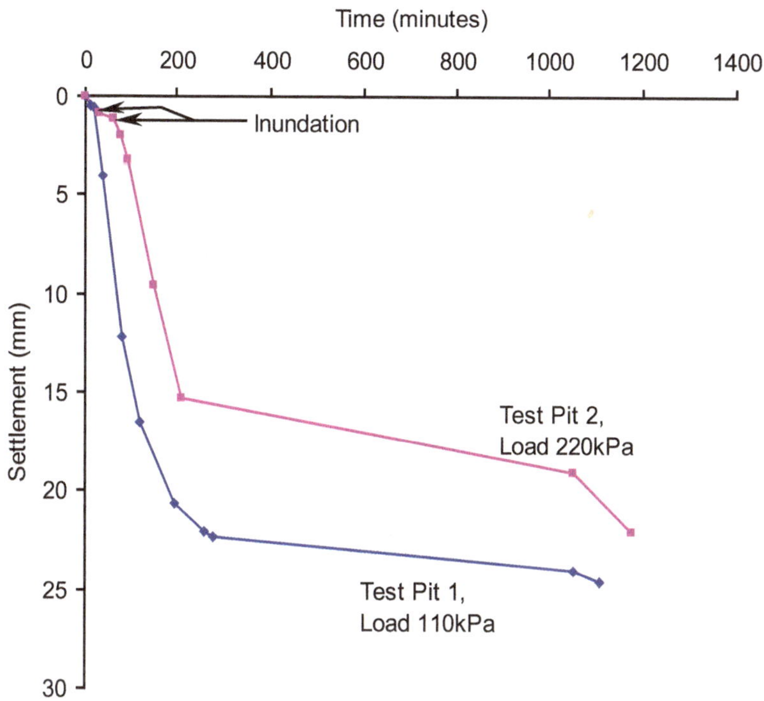

Fig. 5. Plate tests at Taif

In both instances the plates were seen to settle about a millimetre when the ground was dry but only 3 to 4 h after inundation the plates had settled 20 to 25 mm. The range of movement of the gauges were limited so the results were considered in a quantitative rather than strictly qualitative manner.

It had been shown the silts were collapsible. It was also subsequently proven from grading curves that these deposits were poorly graded, which gave rise to high permeability and therefore rapid rates of settlement during the plate tests.

3.4 Building Observations and Recommendations

The main building movements were clearly seen to tie in with the location of vegetation around the building. Movement of the western and southern sectors (Fig. 4) would have been triggered by watering of the planted vegetation, which in turn would have moisten the underlying silts. On the eastern side there was no movement as no vegetation was present.

The main recommendations arising from the investigative work were to:

- Remove any vegetation from around the building and stop watering in those areas
- Place an impermeable membrane around the building and a concrete apron draining away from the structure

4 Conclusions

The foregoing case histories are examples of soil conditions that are regularly encountered in the Middle East. It has been clearly shown that important elements is correctly identifying the presence of collapsible soils are:

- Using the observational skills of experienced geotechnical engineers or engineering geologists, and not just relying on field and laboratory testing.
- Undertaking detailed visual inspection of superficial deposits, preferably using trial pits
- Using plate load tests with inundation to prove possible collapse of any near surface soil
- Recognising that taking undisturbed samples of most collapsible soils is difficult and so laboratory testing may give misleading results

Where construction is contemplated on these soils every endeavour should be made to exclude water infiltration. It is however preferable to place foundations below these deposits, or undertake ground improvement to change the soil structure by such means as dynamic compaction.

Stabilization of Soil Using Terrasil, Zycobond and Cement as Admixtures

V. Padmavathi[1(✉)], E. C. Nirmala Peter[1], P. N. Rao[2], and M. Padmavathi[1]

[1] Department of Civil Engineering, JNTUH College of Engineering, Hyderabad 500 085, India
vpadma70@gmail.com, ecnpjntuh@gmail.com, mpadmace@gmail.com

[2] Department Civil Engineering, BITS Pilani-Hyderabad Campus, Hyderabad, India
pnrao_bits@gmail.com

Abstract. Soil stabilization techniques are aimed at improving the strength and stability of the base soil to meet the requirement of a particular application, and also to increase resistance to softening when saturated or inundated by bonding the particles to yield a durable product. This study presents the use of nano materials like Terrasil, Zycobond and Cement as admixtures to improve the strength properties both at OMC and SMC conditions of a c-ϕ soil (SC) with 31 percentage fines and a plasticity index value of about 15%. The addition of Terrasil to the base soil yielded water tightness to the product and considerable improvement in the strength properties are also observed when Zycobond and Terrasil are used independently as additives and also in combination with cement. Paving blocks of size 19 cm × 23 cm × 10 cm prepared with 3% of cement, Zycobond and Terrasil yielded uniaxial compressive strength of 1.28 MPa after 14 days of curing.

1 Introduction

Soil stabilization technique, used to modify the properties of poor quality soils (Sharma et al. 2012; Abood et al. 2007) available at the site, to make them meet the desired requirements of design/construction. The commonly used materials are fast depleting and hence search for new materials and improved techniques to process the local materials have gained great impetus. Since the nature and properties of the soils vary widely, a suitable stabilization technique has to be used to deal with a particular situation or requirement. Many chemical additives like Olivine – utilized to stabilize soil through Magnesite production during olivine carbonation, which is capable of binding soil particles, SH-85 – a non-traditional stabilizer which is calcium based and contribute to denser soil fabric by filling the pores between the soil particles with a gel formed from the chemical, Silicon dioxide nano particles – which increases the hydraulic conductivity and when used in combination with cement increases the strength (Bahmani et al. 2014). Bioenzymes - chemical, organic and liquid concentrated substance has been used to stabilize soil sub-base courses of pavements (Ravi

H. Shehata and H. Poulos (Eds.): GeoMEast 2018, SUCI, pp. 163–170, 2019.
https://doi.org/10.1007/978-3-030-01917-4_13

Shankar et al. 2009). Additives like lime, cement, flyash, bitumen and nano materials like zycosil, zycobond and terrasil have been used for stabilizing the subgrade soils for road construction. They have improved the CBR values considerably and helped in cost-effective construction of pavements (Adeinola and Nnochiri 2017). In the present study two nano chemicals Terrasil and Zycobond are used as stabilizing agents.

1.1 Terrasil

It is defined as an organosilane compound which reacts with soil particles and converts them (all types of soils) from water loving (Hydrophilic polar) to water hating (Hydrophobic non polar) particles. This makes the soil insensitive to water and can be compacted to give better interlocking to the soil particles. It offers or forms a permanent water repellent nano layer on all types of soils, aggregates etc. This siloxane is non-leachable as it chemically binds to surfaces permanently. It forms strong covalent bond structure allows the treated material to breathe i.e. It allows free flow of air through its structure and preserves thermal insulation property. Terrasil prevents damage due to capillary rise of water, cracking of soil and also resistant to ultraviolet rays. It is highly soluble in water. The chemical action of terrasil (Shirahatti et al. 2016) is shown in Fig. 1.

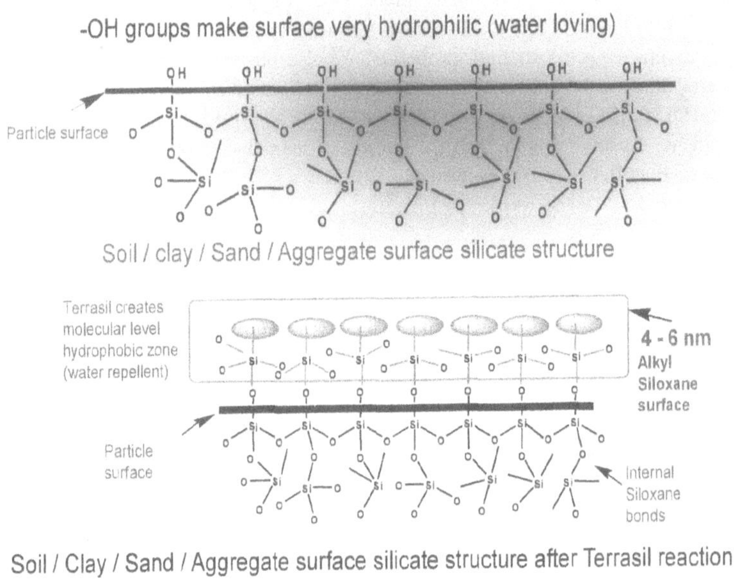

Fig. 1. Chemical reaction of Terrasil

1.2 Zycobond

It is a nano polymer with particle size less than 90 nm. It disperses in the soil bonding the soil particles imparting resistance to erosion, dust control and fatigue. Terrasil mixed with zycobond imparts excellent bonding properties to the soil and provides permanent erosion control. The bonds formed with terrasil are permanent and long lasting and as they resist UV degradation, they remain stable for years.

Construction of landfills require considerable amount of clay for lining and capping the landfills. In zoned embankments also large quantities of clay is required for making the core or hearting zone. In many situations the clay soil may not be available at the construction site and hauling the material from long distances may enormously increase the cost of the project. Reinforced soil foundation beds and RE walls also require granular material or sand to be used for developing good frictional property between the reinforcement and soil and also meet other requirements.

This paper furnishes the details of laboratory investigations made to explore the use of terrasil, zycobond and cement as additives to improve the strength, stability and impervious nature of a locally available soil for the following applications.

- Suitability of the altered material to be used as liner material for landfills to prevent contamination of the surrounding area.
- Stabilize locally available material to be used as core or hearting material
- The thickness of the core can be reduced
- Stabilized local soil to be used in the construction of reinforced earth beds for foundations.
- Terrasil, cement and zycobond form strong bonds – making of paving bricks with locally available soil.

2 Materials and Methodology

A locally available soil with percentage fines of 31 and a plasticity index value of 15% is taken as the base soil for stabilization with different admixtures. The properties of the local soil are furnished in Table 1. The admixtures used are Terrasil, zycobond independently, and combinations of 3%cement + terrasil + zycobond, 6%cement + zycobond and 3% cement + terrasil. Blocks were prepared with 3%cement + terrasil. To check the improvement in the strength and stability of the stabilized soils, shear parameters were obtained from direct shear tests on samples compacted with OMC from standard proctor compaction test. The suitability of the stabilized soil as a liner material and as core material, in addition to the shear properties which reflect the stability, the permeability property was also checked for soils stabilized with zycobond and terrasil independently. The soils are stabilized with additives as follows.

2.1 Soil Stabilized with Only Terrasil

Terrasil solution is prepared in the proportion 1:200. The required quantity of soil (approximately 5 to 6 kg) for conducting direct shear tests and permeability test is

Table 1. Properties of locally available soil

Test conducted		
Grain size distribution		
Gravel (%)	5	
Coarse sand (%)	22	
Medium sand (%)	29	
Fine sand (%)	13	
Silt & Clay (%)	31	
Atterberg limits		
Liquid limit (%)	33	
Plastic limit (%)	18	
Plasticity Index (%)	15	
IS Classification of soil	SC	
Compaction test	*Standard proctor test*	*Modified proctor test*
OMC (%)	11.0	8.0
Maximum dry density (kN/m^3)	18.9	20.6
Direct Shear Tests	*OMC*	*SMC*
Cohesion, c (kN/m^2)	11	07
Angle of internal friction, ϕ ()	38	32
Permeability Test		
Coefficient of permeability, k (cm/s)	9.77×10^{-7}	

taken in a tray and completely inundated with the prepared terrasil solution. The soil is left to air dry for a few days and then oven dried. The OMC obtained for the local base soil is used for the preparation of direct shear and permeability test samples. Direct shear tests were conducted on samples compacted at OMC and also at SMC (saturated samples compacted at OMC).

2.2 Soil Stabilized with Zycobond

For stabilizing the soil with zycobond, the required quantity of soil is taken. Zycobond is added to the optimum moisture content in the proportion 1:99 to prepare the zycobond solution. This water is then added to the soil, and direct shear and permeability tests samples were prepared.

2.3 Soil Stabilized with Terrasil, Zycobond and Cement

Soil is initially treated with terrasil. After oven drying the sample the required quantity of cement is added and OMC water quantity mixed with zycobond is added as explained above. These samples for direct shear are cured for 7 days before conducting the tests. The shear parameters from direct shear tests are furnished in Table 2. The variation in shear parameters is shown in Figs. 2 and 3.

Table 2. Shear parameters from direct shear test on local soil stabilized with different admixtures

Shear parameters	Local soil	Local soil with zycobond	Local soil with Terrasil	Local soil with 3% cement, Terrasil and zycobond	Local soil with 6% cement and zycobond
@ OMC					
Cohesion, c (kN/m^2)	11	07	05	–	–
Friction angle, ϕ (°)	38	35	46.8	–	–
@ SMC					
Cohesion, c (kN/m^2)	07	05	0	35	20
Friction angle, ϕ (°)	32	34	47.3	53	44

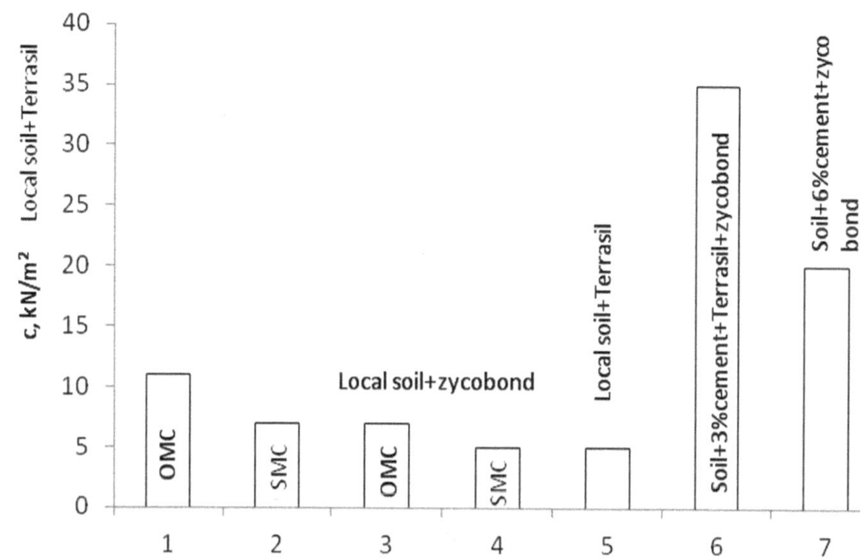

Fig. 2. Variation of c (kN/m^2) for different admixtures

Wherever Terrasil is used, the soil is initially treated with terrasil as described above, and after oven drying the soil sample the other admixtures are used on the soil. Wherever cement is used, minimum curing periods of 14 days are allowed before conducting the tests on the treated soil. Permeability tests are conducted only for three soils, local base soil, soil treated with zycobond and soil treated with terrasil. The permeability values are shown in Table 3.

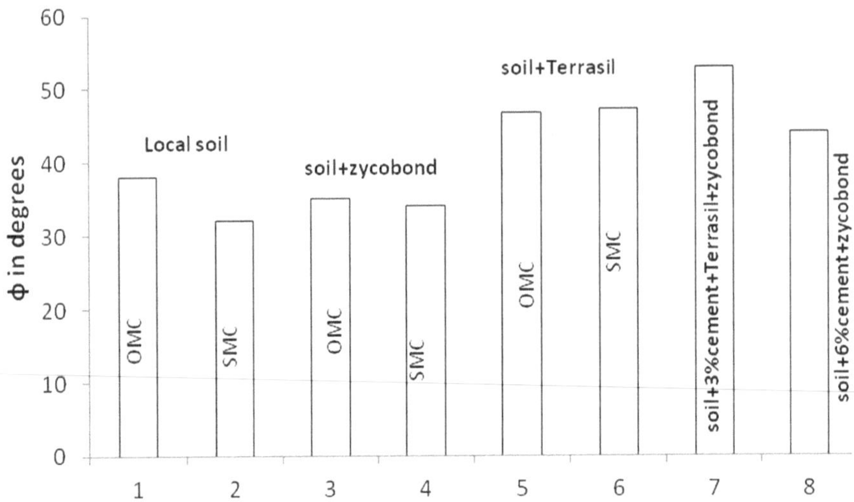

Fig. 3. Variation of ϕ for different admixtures

Table 3. Permeability tests on local and treated soils

Soil type	Local soil	Soil treated with zycobond	Soil treated with Terrasil
Coefficient of permeability, k (cm/s)	9.77×10^{-7}	1.26×10^{-6}	Impermeable

Paving blocks are prepared with local soil, 3% cement and terrasil. The size of the block is 23 cm × 19 cm × 10 cm as shown in Fig. 4. Uniaxial compression test was carried out on these blocks after curing them for a period of 14 days. The tests yielded an average compressive strength of about 1.28 MPa.

3 Results and Discussions

The shear tests conducted on the locally available base soil have shown that on saturation the cohesion and angle of internal friction decreased considerably. However when soils are treated/stabilized with zycobond, both OMC and SMC almost the same parameters were obtained. This indicates that there is no effect of saturation on the shear parameters and they not much different from base soil properties. This indicates that when only zycobond is used as an admixture, there is improvement in the strength properties of the soil but the soil becomes stable without being affected by moisture content.

In case of terrasil variation is obtained for bonding property (cohesion) and it has become zero on saturation, but there is no variation in the angle of internal friction for OMC and SMC. However the friction values are increased by almost 8 to 9 degrees when compared with the base soil, indicating considerable increase in the strength and

Fig. 4. Paving block

stability of the treated soil. For the remaining trials with combination of admixtures tests were conducted only at SMC. When both zycobond and terrasil are added with 3% of cement, there is considerable increase in both the shear parameters demonstrating maximum benefit interms of strength and stability can be obtained for this combination of admixtures. Soil mixed with 6% of cement and zycobond has also shown good increase in cohesion and angle of internal friction. So depending on the requirement this can also conveniently used for improving the strength properties of the local soil.

Permeability tests yielded complete imperviousness of the treated material when terrasil is used as an admixture. Treatment with only zycobond resulted in slight increase in the permeability of the material.

4 Conclusions

1. Keeping in view the properties of terrasil such as forming strong permanent bonds with soil particles, imparting strength and imperviousness to the material
   (a) terrasil treated local soils can be used as replacement material for clay in liners for landfills and also for capping purpose. However in case of industrial waste landfill, based on the leachate that gets generated, additional investigations have to be carried to check whether the leachate has any detrimental effect on the stabilized soil.
   (b) Terrasil stabilized local soil can also be used as a replacement material for clay as core or hearting material in zoned earthen embankments and further the thickness of the zoned section can also be reduced.

2. When terrasil is used as an admixture to the local soil, there is considerable increase in the friction property of the modified material, imparts resistance to UV rays, makes the matcrial imperviousness – therefore if used for reinforced earth beds, it increases the life and also increases the strength of the soil. However this aspect has to be further explored.
3. Blocks prepared from terrasil and cement yielded good compressive strength. Hence they can be used in the preparation of paving blocks.
4. Combination of terrasil and cement yielded the best of the results – hence can be beneficially used for treating the local soils.

References

Bahmani, S.H., Huat, B.B.K., Asadi, A., Farzadnia, N.: Stabilization of residual soil using SiO_2 nano particles and cement. Constr. Build. Mater. **64**, 350–359 (2014)

Ravi Shankar, A.U., Rai, H.K., Mithanthaya, R.: Bioenzyme stabilized lateritic soil as a highway material. J. Indian Road Congr. (2009)

Adeinola, O.S., Nnochiri, E.S.: Stabilizing lateritic soil using terrasil solution. SSP J. Civ. Eng. **12**(1), 19–27 (2017)

Shirahatti, P.N., Kallimani, P.M., Rajashekhar, M.S.: Application of geo-textile and terrasil chemical to reduce permeability of soil. IJSRD **4**(6) (2016)

Abood, T.T., Kasa, A.B., Chik, Z.B.: Stabilization of silty clay using chloride compounds. J. Eng. Sci. Technol. **2**(1), 102–110 (2007)

Sharma, N.K., Swain, S.K., Sahoo, U.C.: Stabilization of a clayey soil with flyash and lime: a micro level investigation. Geotech. Geol. Eng. **30**, 1197–1205 (2012)

Experimental Investigation on Stone Columns in Cohesive Soil

Adel Hanna[1], Mahmoud Khalifa[1(✉)], and Mohamed Abdel Rahman[2]

[1] Department of Building, Civil and Environmental Engineering, Concordia University, 1455 De Maisonneuve Blvd. W., Montréal, QC, Canada
mahm_kh@encs.concordia.ca

[2] Housing and Building National Research Center, 87 El-Tahreer St, El-Doqi, Giza Governorate, Egypt

Abstract. Foundations on soft cohesive soil suffer from excessive settlement and low bearing capacity. In practice, vibro-displacement stone columns are widely used as cost-effective and environmentally friendly soil improvement technique. The high strength and stiffness of stone columns enhance the bearing capacity and reduce the settlement of the foundations.

The parameters governing the performance of these columns are geometry (column length, diameter, and spacing), stiffness of stone columns material, and properties of surrounding soil. This paper presents a new sophisticated laboratory testing technique to examine the performance of clay soil reinforced by single and groups of stone columns. In this technique, for the first time, the effect of the compaction energy utilized during stone columns installation process was examined. In addition, the influence of spacing between stone columns on the shearing resistance of reinforced clay soil was investigated.

It was concluded that increasing the compaction energy improves the performance of stone columns at the same replacement ratio. It is of interest to know that single stone column exhibited better capacity as if it was installed in a group of columns, particularly for high replacement ratio. Also, increasing the spacing between stone columns reduces the total shear resistance of the reinforced soil.

1 Introduction

The concept of stone columns is stem from the fact that by adding compacted granular material to the soft cohesive soil will improve its overall stiffness, and accordingly its strength. Accordingly, the bearing capacity of the soil increases and the corresponding settlement reduces (Etezad et al. 2015; Kelly 2014; Mitchell and Huber 1985; Muir Wood et al. 2000; Priebe 1995). Furthermore, stone columns reduce the drainage path of the water during consolidation due to its high permeability comparing to the native clay soil (Sivakumar et al. 2004). Moreover, stone columns are used to increase the stability of natural slopes and reduce liquefaction potential when used in loose cohesionless soil.

Stone columns are installed by either replacement method, which by making a hole in the ground, then fill it up with stones, which is suitable for sensitive clay, or the

H. Shehata and H. Poulos (Eds.): GeoMEast 2018, SUCI, pp. 171–181, 2019.
https://doi.org/10.1007/978-3-030-01917-4_14

displacement method, which is known as Vibro-replacement method (Baumann and Bauer 1974). In the Vibro-replacement method, a vibrator sinks in the clay soil under its weight, assisted by air jet for a partially saturated soil, or water jet in saturated soil, to the required depth, while granular materials are gradually fill in the hole, well-graded gravel of average diameter between 25 mm and 50 mm is used to make stone columns (Muir Wood et al. 2000). Stone columns are installed in groups with a triangular, hexagonal, or square arrangement, which replaces 10–35% of the surface area of the cohesive soil (Hu 1995; Black et al. 2007).

The parameters govern the performance of soft soil reinforced by stone columns are mainly the geometric properties of the columns (length, diameter, and spacing), and the stiffness of their material, and the native surrounding soil. Unit cell concept was used to estimate the bearing capacity of the ground reinforced with stone columns (Hughes and Withers 1974; Gibson and Anderson 1961). The method adopt the bulging mode of failure as the only possible failure mechanism, which neglected columns interaction and group performance (Balaam and Booker 1981). The group interaction reported by Hu (1995) was confirmed by the numerical analysis carried out by Lee and Pande (1998); Muir Wood et al. (2000). Hanna et al. (2013) reported that stone columns may fail in general, local or punching shear failure. They develop a theory to predict the mode of failure for a given geometry/ground conditions.

Bouassida et al. (2009) presented design charts to determine the ultimate bearing capacity of a group of floating stone columns. However, they neglected the friction between the footing and the soil and the effect of the stone columns arrangement. Etezad et al. (2015) have developed a theory for bearing capacity of reinforced ground with stone columns using limit equilibrium method and equivalent soil properties under the footing. However, the model did not include other possible modes of failure. Khalifa et al. (2017) presented an analytical model using the method of slices to estimate the bearing capacity of soil reinforced by stone columns. They concluded that the bearing capacity of reinforced soil reduces due to the increase of the spacing between stone columns.

2 Experimental Work

Creating a prototype simulation of a clay soil reinforced with a group of stone columns is not a simple process. In the presented paper a new experimental technique was used to investigate the influences of the applied compaction energy during stone columns installation and spacing between stone columns on the performance of the reinforced clay soil. In this technique, the complexity of Vibro floatation method used for stone columns installation was simplified to a displacement method, and the non-homogeneity in the field layers was simplified to a one-dimensionally consolidated bed of homogenous clay. Clay sample preparation and stone columns installation process were conducted in a large rigid side wall tank (Fig. 1). In order to reduce the scale and boundary effects, nine 10 cm diameter core samples (CTS) of the reinforced clay were extracted from the tank and tested in a triaxial system.

The test program consists of creating two large clay samples (T1 & T2), each subjected to the same preconsolidation pressure, but different compaction energy was

Fig. 1. Consolidation and stone columns installation setup.

used during stone columns installation. The utilized compaction energy and preconsolidation pressure are shown in Table 1. Kaolin clay powder was used for creating the clay samples, due to its low sensitivity, high consolidation rate, and it can be saturated at a relatively short time. The dry powder was mixed with water at 1.5 of its liquid limit (*LL*). The liquid limit of the kaolin clay was 60% thus weight of water added to dry powder was 90% of its dry weight. The slurry was mixed using a mixer, for 30 min until a well-mixed slurry was obtained. A sand bed of 5 cm was placed below and above the Kaolin slurry layer as shown in Fig. 2. Two porous sheets were sited between the sand layers and the Kaolin slurry. The friction between the Kaolin slurry and the tank walls has a marked influence on transferring stress to the bottom of the clay layer. This friction causes a reduction in the bottom of the tank up to 20% of the applied pressure on the top of the tank (Cairncross 1973). In order to reduce the friction between the Kaolin slurry and the tank walls, all the sides of the tank are smeared with silicone grease.

Table 1. Large clay samples properties

Test no.	Compaction energy (kN.m/m^3)	Preconsolidation pressure, σ_c (kN/m^2)
T1	310	80
T2	620	80

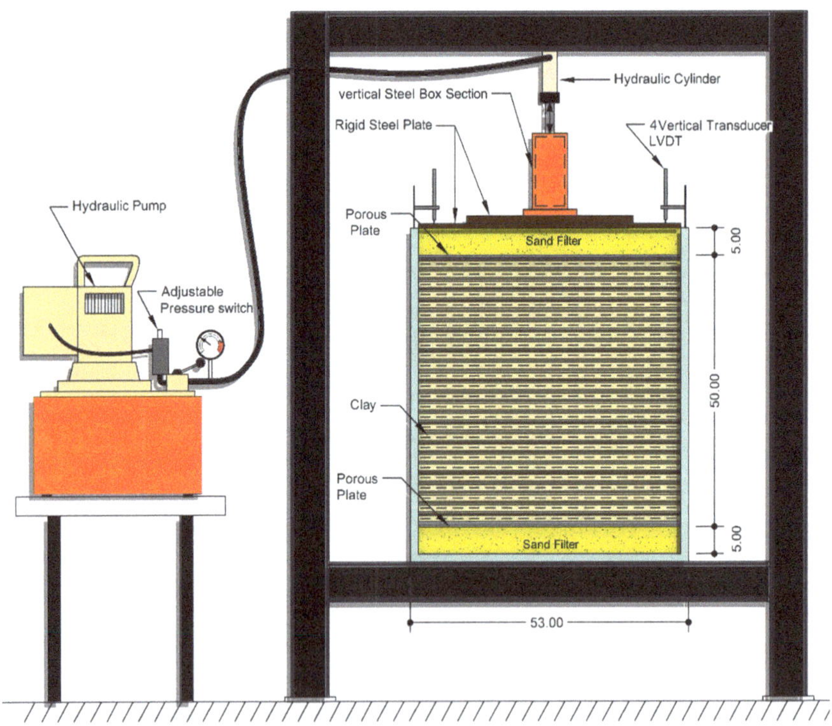

Fig. 2. Large clay sample consolidation setup

After filling the tank, the consolidation process was started by gradually increasing the applied pressure on the top of the soil. The pressure was applied in increments, each increment lasted until steady state was reached. Measuring the pore water pressure in the clay slurry was not possible. Accordingly, the rate of the settlement was measured to judge the steady state. The initial applied stress is 20 kPa, then it was doubled every 24 h until reaching the maximum consolidation pressure of 80 kPa, which was remained for four days before removing the applied pressure.

The displacement stone column installation method was developed by driving a tube closed with a separate cone to a depth of 30 cm. The tube is gradually pulled out, and the sand was then filled in (Fig. 3). The method of compaction has a significant effect on the behavior of stone columns (Miranda et al. 2015; Bergado and Lam 1987). Thus, two compaction energies were used, namely 311 kN.m/m^3 for tests no. (T1) and 623.6 kN.m/m^3 for tests no. (T2). The values of the compaction energy were selected to achieve two criteria; keeping the sand in place after removing samples from the tank to perform the triaxial test, and minimize the change in the stone column replacement ratio due to compaction. The sand inside the columns was compacted by a weight of 300gm freely drop from a constant height. The details of the compaction properties are shown in Table 2. Stone columns were constructed in the large tank with different arrangements as shown in Fig. 4.

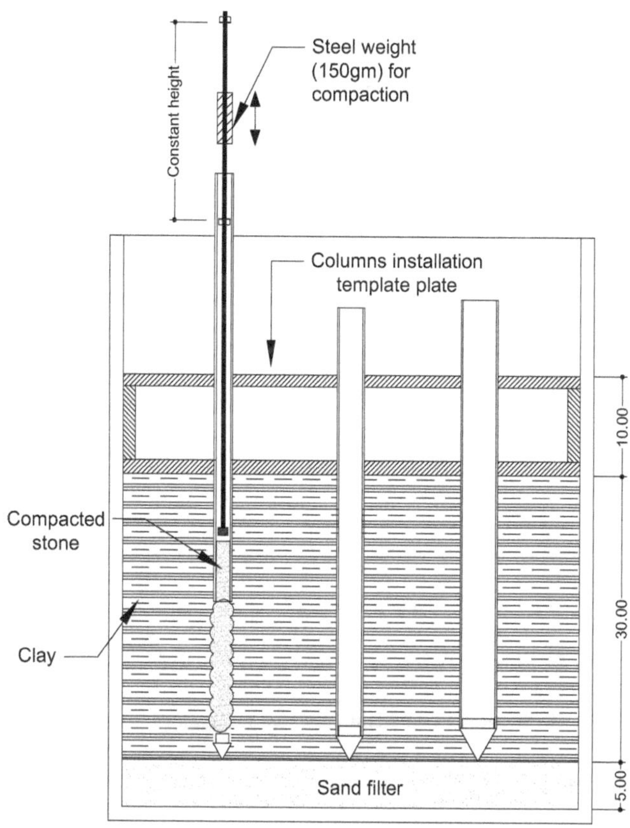

Fig. 3. Stone columns installation method

Table 2. Compaction properties for stone columns

Stone columns diameter (cm)	Volume of stone in each layer (cm^3)	Height of drop (cm)	No. of drops	
			T1	T2
2.1	13.8	24	6	12
3.0	28.3	24	12	24
3.9	47.8	24	20	41
5.1	81.7	24	35	70

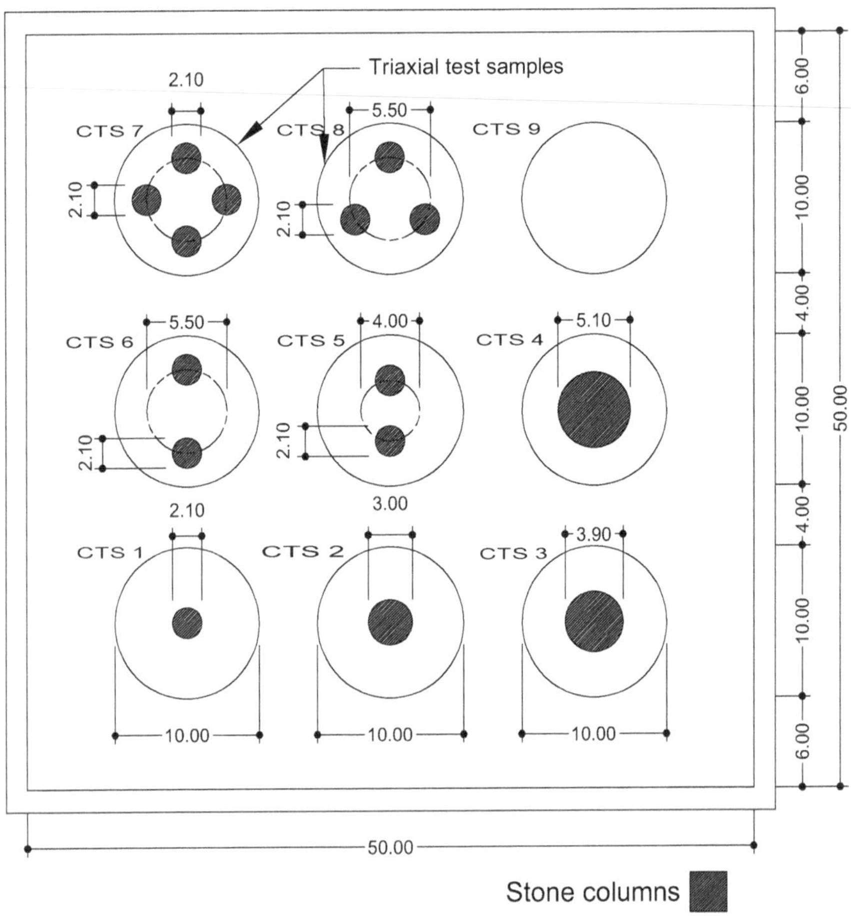

Fig. 4. Stone columns and core triaxial samples arrangement

Nine core triaxial samples (*CTS*) were extracted from the clay blocks using thin-wall cutter cylinders as shown in Fig. 5. In order to remove the clay around the cutters, tank sides were disassembled, and the clay around the cutter cylinders was carefully scraped. Stone columns properties in each CTS are shown in Table 3, where (*S*) and

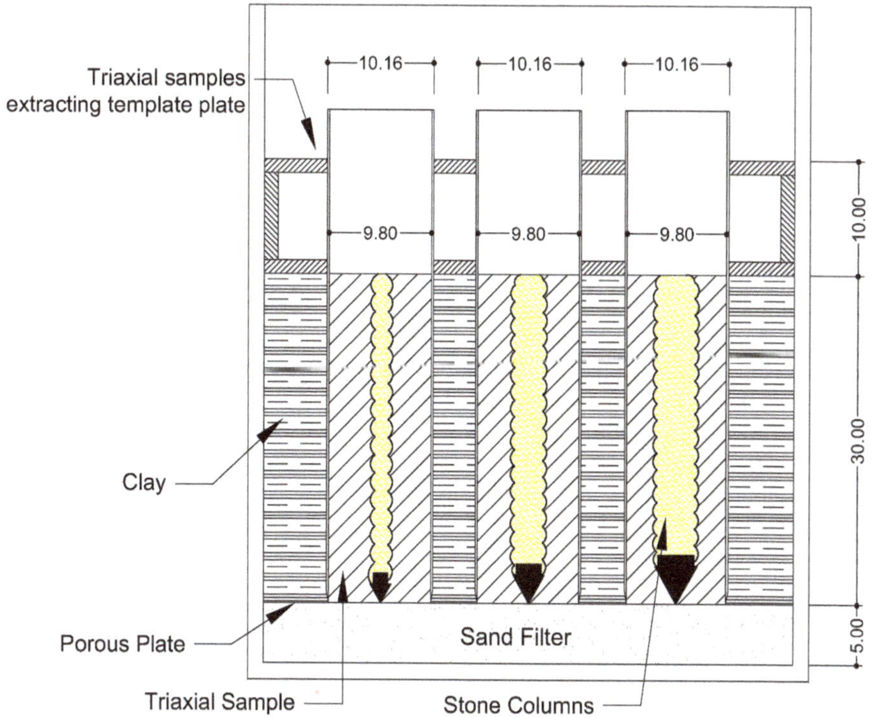

Fig. 5. Triaxial samples cutting process

(*D*) are the spacing and the diameter of stone columns, (*As*%) is the replacement ratio, which is calculated as the cross-section area of stone columns divided by the total cross-section of the composite sample.

Consolidated undrained (*CU*) triaxial test was carried out on each core sample. The sample was placed in the triaxial cell according to ASTM (2004), D 4767-04 standard. A confinement pressure (σ_3) 160 kPa was applied to the sample. The consolidation stage was allowed for 12 h to achieve 100% of primary consolidation; then the loading was applied to the sample using strain control method.

Table 3. Core triaxial samples (CTS) properties

Sample no.	No. of stone columns	Stone columns diameter, *D* (cm)	Stone columns spacing ratio, *S/D*	Replacement ratio, *As* (%)
CTS 1	1	2.10	–	4.4
CTS 2	1	3.00	–	9.0
CTS 3	1	3.90	–	15.2
CTS 4	1	5.10	–	26.0
CTS 5	2	2.10	2.0	8.8
CTS 6	2	2.10	2.75	8.8
CTS 7	4	2.10	2.0	17.6
CTS 8	3	2.10	2.0	13.2
CTS 9	–	–	–	0

3 Test Results

The data obtained from triaxial tests was analyzed assuming that stone columns and surrounding clay soil behave as one homogeneous soil. So, the deviator stress (q) is the axial load divided by the total cross-section of the sample and the axial strain (ε) is the axial deformation divided by the total length of the sample. The stress-strain curves for samples CTS 1 & 5 & 7 & 8 are presented for tests T1 & T1 in Fig. 6. There is a slight increase in the deviatoric stress for the reinforced soil due to increasing compaction energy in test T2. The compaction was applied on stone columns. So, increasing the compaction energy led to an increase in the relative density of stone inside stone columns. Accordingly, the overall resistance of reinforced clay samples increased. However, the effect of compaction is noticeable in CTS 7 & 8 which have a higher replacement ratio, 18%, and 13% respectively.

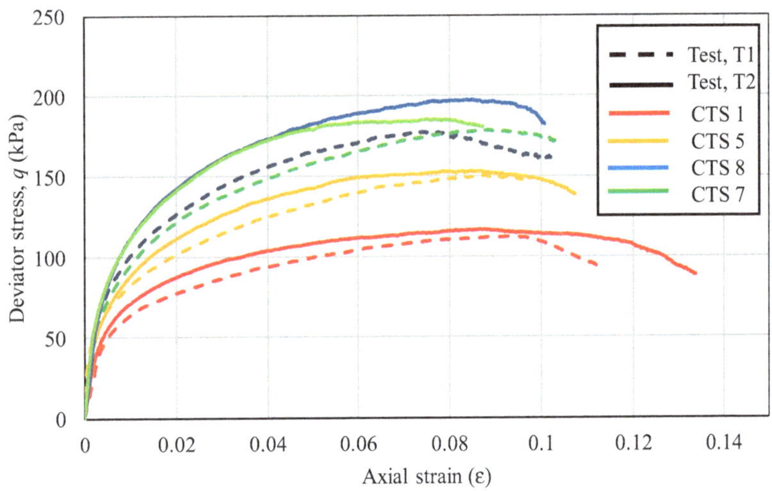

Fig. 6. Stress-strain curve for Test T1 and T2

In order to investigate the difference between single and group of stone columns on the performance of reinforced clay soil, the deviatoric stress at failure (q_f) for all reinforced samples is shown in Fig. 7. For samples reinforced by a single stone column, the bulging failure mode was only observed. So, there is a linear relationship between the replacement ratio (As) and maximum deviator stress. Nevertheless, two modes of failure were observed in case of group of stone columns namely; shear and bending. For small replacement ratio ($As < 15\%$), the tone columns were closer to the center of the sample and shear failure was observed. However, for samples replacement ratio higher than 15% stone columns were closer to the edge of samples as shown in Fig. 4, and bending failure was observed. Therefore, the relationship is not linear for clay samples reinforced by a group of columns. For replacement ratio less than 15%, the capacity of the group slightly increased comparing with single stone column at the

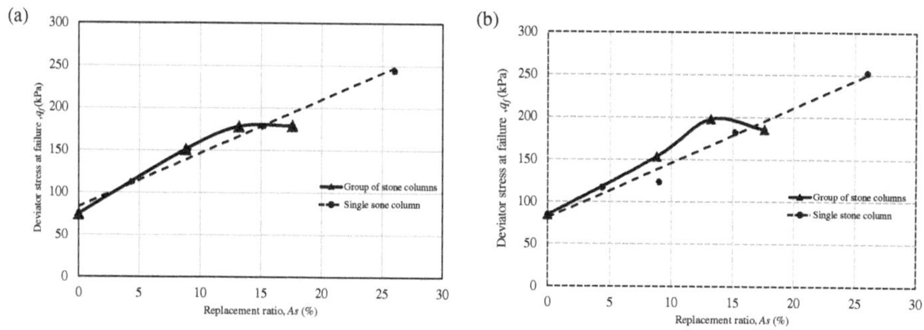

Fig. 7. Deviator stress at failure for single and group of stone columns, (a) test T1; (b) test T2

same replacement ratio. However, the resistance of the group significantly reduced for replacement ratio higher than 15%. The same behavior was observed for T1 and T2.

Figure 8 presents the stress-strain behavior for samples *CTS 5* and *CTS 6* which have the same replacement ratio (*As*) and (*S*/*D*) changed from 2 to 2.75 respectively. It can be noted that increasing the spacing between columns at the same replacement ratio reduces the resistance of the reinforced soil, which agreed well with (Castro 2014; Khalifa et al. 2017) observations. The resistance reduces with the increase of the columns spacing ratio (S/D) which confirmed that for small spacing between columns the lateral support from the surrounding soil increases and accordingly, shows improvement.

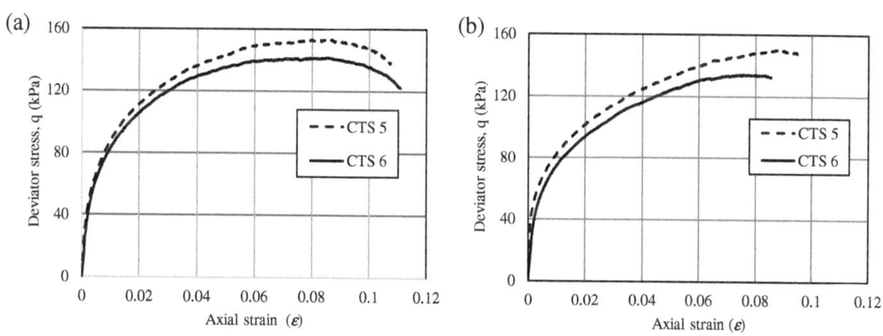

Fig. 8. Stress-strain response under undrained loading for group of 2 columns with *As* = 8.8%, (a) test T1; (b) test T2

4 Conclusions

Consolidated undrained triaxial tests were used to investigate the performance of clay soil reinforced by single and group of stone columns. A new technique was developed to install stone columns in the lab, which simulate the field installation method. From the CU triaxial test result, the following was concluded:

1. Increasing the compaction energy during stone columns installation process is more efficient for high replacement ratios.
2. For single stone column, the deviator stress linearly increased with the increase of the replacement ratio.
3. The performance of singles and group of stone columns is similar for replacement ratio less than 15%.
4. For replacement ratio more than 15%, there is a significant reduction in the shearing resistance for the group of stone columns as compared to a single column at the same replacement ratio.
5. Increasing S/D ratio lead to a slight reduction in the shear resistance of clay soil reinforced by a group of stone columns at the same replacement ratio.

Acknowledgments. The financial support received from National research council of Canada and Concordia University are acknowledged.

References

ASTM: D 4767-04: Standard test method for consolidated undrained triaxial compression test for cohesive soils. ASTM Int., West Conshohocken (2004)

Balaam, N., Booker, J.R.: Analysis of rigid rafts supported by granular piles. Int. J. Numer. Anal. Methods Geomech. **5**(4), 379–403 (1981)

Baumann, V., Bauer, G.: The performance of foundations on various soils stabilized by the vibro-compaction method. Can. Geotech. J. **11**(4), 509–530 (1974)

Bergado, D.T., Lam, F.L.: Full scale load test of granular piles with different densities and different proportions of gravel and sand on soft Bangkok clay. Soils Found. **27**(1), 86–93 (1987)

Black, J., Sivakumar, V., McKinley, J.: Performance of clay samples reinforced with vertical granular columns. Can. Geotech. J. **44**(1), 89–95 (2007)

Bouassida, M., Jellali, B., Porbaha, A.: Limit analysis of rigid foundations on floating columns. Int. J. Geomech. **9**(3), 89–101 (2009)

Cairncross, A.M.: Deformations Around Model Tunnels in Stiff Clay. University of Cambridge, Cambridge (1973)

Castro, J.: Numerical modelling of stone columns beneath a rigid footing. Comput. Geotech. **60**, 77–87 (2014)

Etezad, M., Hanna, A., Ayadat, T.: Bearing capacity of a group of stone columns in soft soil. Int. J. Geomech. **15**(2), 04014043 (2015)

Gibson, R., Anderson, W.: In situ measurement of soil properties with the pressuremeter. Civ. Eng. Public Works Rev. **56**(658), 615–618 (1961)

Hanna, A., Etezad, M., Ayadat, T.: Mode of failure of a group of stone columns in soft soil. Int. J. Geomech. **13**(1), 87–96 (2013)

Hu, W.: Physical modelling of group behaviour of stone column foundation. Ph.D. dissertation, University of Glasgow, UK (1995)

Hughes, J., Withers, N.: Reinforcing of soft cohesive soils with stone columns. Ground Eng. **7**(3), 42–49 (1974)

Kelly, P.: Soil Structure Interaction and Group Mechanics of Vibrated Stone Column Foundations. University of Sheffield, Sheffield (2014)

Khalifa, M., Etezad, M., Hanna, A., Sabry, M.: Bearing capacity of strip foundation on soft soil reinforced with stone columns using method of slices. Paper presented at the International Congress and Exhibition "Sustainable Civil Infrastructures: Innovative Infrastructure Geotechnology" (2017)

Lee, J.S., Pande, G.N.: Analysis of stone-column reinforced foundations. Int. J. Numer. Anal. Methods Geomech. **22**(12), 1001–1020 (1998)

Miranda, M., Da Costa, A., Castro, J., Sagaseta, C.: Influence of gravel density in the behaviour of soft soils improved with stone columns. Can. Geotech. J. **52**(12), 1968–1980 (2015)

Mitchell, J.K., Huber, T.R.: Performance of a stone column foundation. J. Geotech. Eng. **111**(2), 205–223 (1985)

Muir Wood, D., Hu, W., Nash, D.: Group effects in stone column foundations: model tests. Geotechnique **50**(6), 689–698 (2000)

Priebe, H.J.: The design of vibro replacement. Ground Eng. **28**(10), 31 (1995)

Sivakumar, V., McKelvey, D., Graham, J., Hughes, D.: Triaxial tests on model sand columns in clay. Can. Geotech. J. **41**(2), 299–312 (2004)

Experimental and Numerical Study of the Impact of Heavy Structures Settlement Founded on Layered Soft Ground Improved by Floating Rigid Inclusions

A. Rouag[1], M. Akchiche[2(✉)], and J. Moussa[1]

[1] Université Saad Dahlab Blida1, Blida, Algeria
[2] Faculty of Civil Engineering, USTHB, Bab-Ezzouar, Algeria
makchiche@usthb.dz

Abstract. This article treats experience feedback of the behavior of six tanks founded on layered soft ground reinforced by rigid inclusion "RI" columns of floating "INSER" in the plain of Mitidja, Boufarik, Algeria. The rigid inclusions columns have a diameter 0.42 m and 26.3 m to 28.3 m length, uniformly distributed according to a grid of $1.8 \times 1.8\ m^2$. The load transfer mattress would was 2.30 m in thickness and consist of densely rolled granular material with geogrid depending on spacing of the stiff columns. After the Achievements of the realization of the works a loading is carried out on the latter. Settlements appeared exceed even settlements acceptable of work "15 cm" by thus compromising the project. The instrumentation implemented allowed the follow-up uninterrupted settlements according to the evolution of the stresses under the foundation rafts. Numerical calculations by finite elements and analytical methods were carried out and the results obtained are compared with measurements.

1 Introduction

Urbanization is occurring at a furious rate in Algeria specifically in Algiers city, this is mainly due to human migration from rural areas to urban cities. Consequently, areas exhibiting weak soil strata which were previously considered unsuitable for construction are being developed to meet the high demand in service facilities such as road embankment airfield, residential and light commercial and industrial structures. Although pile foundations may be deployed in those areas to meet the design requirements for stability, integrity and serviceability of the infrastructures, they are costly.

Rigid Inclusion is a proven technology in improving the soft ground for many geotechnical applications. Rigid Inclusion are normally constructed to penetrate soft soil layer and founded on more competent soil layer. This is termed as fully penetrating columns or end bearing columns (Boussetta et al. 2016, Briançon et al. 2015). Nonetheless, partially penetrating columns or floating columns with toe embedded within soft clay layer are sometimes used.

H. Shehata and H. Poulos (Eds.): GeoMEast 2018, SUCI, pp. 182–192, 2019.
https://doi.org/10.1007/978-3-030-01917-4_15

2 Dimension of Structures Constructions

The project concerns the construction of the power station at BOUFARIK (Algeria). Reinforcement in RI. Structures built on rigid inclusions are Fuel Oil Tanks (n°5), Demi-Water Tanks (n°6) and Service Water Tanks (n°7) (Fig. 1). The buildings are built on a raft fondations base, the dimensions of these supports are as Table 1.

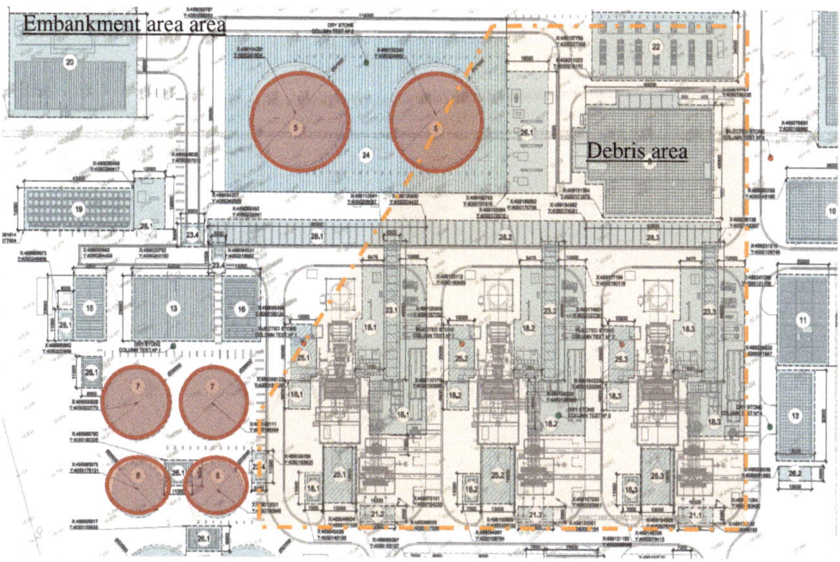

Fig. 1. Arrangement of the works on the Boufarik power station

3 Rigid Inclusion Construction Methods

Rigid Inclusions are a method of reinforcing organic or very weak soils with a structural column to reduce settlements. The column was installed by rotary methods such as displacement auger "INSER"(Fig. 2).

During the descent of the tool, the soil that rises through the lower helical blade is discharged by the smooth central portion (Fig. 3). Conversely, during the ascent, the direction of rotation remains the same and the soil is repressed this time by the upper blades and the truncated central portion.

The depth of rigid inclusions of type "INSER" is presented in the Table 2:

For structures 5, 6 and 7 in rigid inclusion, it has a load transfer mattress which has a thickness of 2.30 m used for the central inclusions (excluding concrete ring device) and 0.50 m for rigid inclusions under the peripheral ring. The mattress is made up of a highly rolled granular material with a dependent geogrid.

The compacting objectives for the different layers are recalled as follows

- The working platform in the quotation 17,60NGA, must be compacted and leveled, to obtain an EV2 $\geq$ 50 MPa

Table 1. Size of structures

Number	*Structure*	Size of the slab R (radius in m)
5	Fuel Oil Tanks	37
6	Demi-Water Tanks	32
7	Service Water Tanks	26

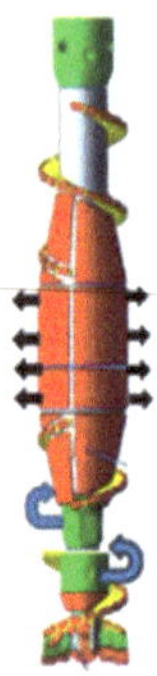

Fig. 2. Principle of soil displacement by the tool INSER®

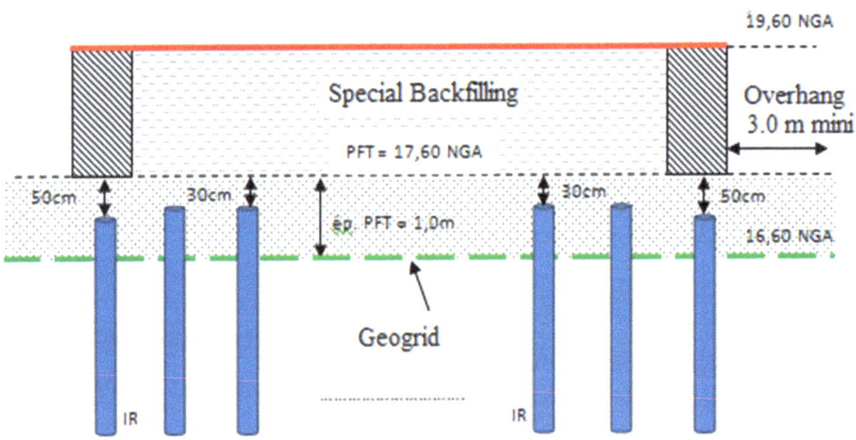

Fig. 3. Arrangement of rigid inclusions

Table 2. Length of rigid inclusions

Structures number	Natural ground level [NGA]	Base level of raft foundation (e = 2 m) [NGA]	Base foundation strengthening [NGA]
5	18.70	17.60	− 11.00
6	18.70	17.60	− 11.00
7	18.60	17.60	− 9.00

- Specific backfill «Special Backfilling» must be compacted to 19,60NGA for EV2 $\geq$ 100 MPa

The INSER rigid inclusions have a diameter of 0.42 m.

The rigid inclusions consist of a concrete C25/30 with the compressive resistance of which jn 28 days is equal to 25 MPa. The concrete is characterized by a long-term deformation modulus of 9320 MPa. The Rigid inclusions are installed in a square pattern with spacing equal to 1.80 m.

4 Nature and Characteristics of the Soils

The site was the object of several geotechnical investigation campaign, we will be limited to the tests which enter to our calculations

The surveys highlighted successively the following geological formations

- Ocher brown silts and clays moderately firm: over thickness varying from 4 to 6 m (formation 1 and 3).
- Soft blackish gray clays 2 to 4 m thick (formation 2),
- Clay alluvium (formation 4 to 5 according to their compactness), with sandy passages between 8 and 14 m and also deeper between 20 and 26 m approximately.

4.1 Compressibility of the Grounds

The results of the tests laboratory concerning the state and the compressibility of the grounds are recapitulated in the following Table 3.

Table 3. Median value retained for the test Oedometric

Formations	Cc	Cg	OCR	qc (MPa)
Brown ochre (ocher)	0.147	0.0147	2.2	1.0
Black clay hearth	0.216	0.0216	1	0.5
Gray clay	0.151	0.0151	1	1.5
Gray clay with sandy loam	0.104	0.0104	1	2.4
Gray clay	0.135	0.0135	0.5	1.1

According to the correlations of Vidalie 1977, the values of Cc seem coherent. On the other hand the reports of “Cc/Cs” between 2 and 5 far away from the ratio of 10 are usually noted. The report of over-consolidated “OCR” means indicates an over-consolidated surface formation (formation 1) and probably formation 3. On the other hand, formation 2 is doubtless normally consolidated in spite of an OCR means higher than 1 just as the argillaceous formations 4 and 5 deep.

The geotechnical synthesis for our calculations is recapitulated in the Table 4. The thicknesses of the various formations are defined by the “CPT” and the characteristics of compressibility are defined by the tests laboratory.

Table 4. General characteristics of the ground snuff for calculation.

Materials	Base rating of the layer (NGA)	Qc MPa	N_{160}	Cc	Cs[a]	OCR	Cc/(1 + e0)	Cs/(1 + eo)
Brown ochre (ocher)	15.0	1.0	5	0.147	0.0147	2.2	0.0783	0.00783
Black clay hearth	12.0	0.5	5	0.216	0.0216	1	0.109	0.0109
Gray clay	7.0	1.5	15	0.151	0.0151	1	0.0857	0.00857
Gray clay with sandy loam	1.0	2.4	15	0.104	0.0104	1	0.0595	0.00595
Gray clay	< −6.0	1.1	10	0.135	0.0135	0.5	0.0801	0.00801
Gray clay		–	30	–	–	–	–	–

[a]The Cs takes into account the ratio of 10 between Cc/Cs.

The comparison of these different values (qc, NSPT and lab tests) shows a certain coherence in the definition of the compressibility of the different geological formations. This consistency allows us to favor the results from the static penetrometer which have the advantage of providing a continuous profile and underscore the heterogeneities of the site (Table 5).

Table 5. Synthesis structures number 5, 6 and 7.

Materials	Structures number	Base rating of the layer (NGA)	qc (MPa)	α_c *	E [MPa]
Brown ochre (ocher)	5	13.5	1.2	6	7.2
	6	15.0	1.0	6	6.0
	7	13.5	1.3	6	7.8
Black clay hearth	5	12.5	0.6	6	3.6
	6	13.5	0.5	6	3.0
	7	12.5	0.5	6	3.0
Gray clay	5	9.0	1.0	6	6.0
	6	9.0	1.0	6	6.0
	7	8.0	1.2	6	7.2
Gray clay with sandy loam	5	2.0	4.0	4	16.0
	6	−1.0	3.5	4	14.0
	7	0.5	4.0	4	16.0
Gray clay	5	<2.0	1.5	6	9.0
	6	<−1.0	2.0	6	12.0
	7	<0.5	2.0	6	12.0

4.2 Hydraulic Conditions

During the reconnaissance campaign, the highest measured water level was at the natural terrain level between 18 and 20 NGA.

5 Analytical Calculation

The most important parameter for characterizing the efficacy of treatment is the settlement reduction factor β given by

$$\beta = \frac{s_i}{s_f} \tag{1}$$

s_i: Settlement of unreinforced soil
s_f: Settlement after reinforcement

The reduction factor of settlement of IR structures is represented in the Table 6.

Table 6. Settlement reduction factor

Structures	Settlement of unreinforced soil (cm)	Settlement after reinforcement(cm)	β
5	46	10.5	4.4
6	58	14.5	4.0
7	49	14.2	3.45

6 Numerical Simulations

The data set for rigid inclusions is presented in Table 7. The geometric characteristics of the inclusions as well as the characteristics of the soil are as defined previously. The soil is modeled in Mohr-Coulomb except the layers 5 and 6 (under the rigid inclusions) in Hardening Soil Model "HSM" to make it possible to take account of the soil being consolidated, we suspect this layer under-consolidated (OCR = 0.5) to generate significant long-term settlements.

An axisymmetric analysis has been carried out using the elastic-perfectly plastic Mohr–Coulomb failure criterion for soft clay, stiff clay. All the analyses have been carried out in the drained condition. The gravity loading has been considered to generate the initial stress conditions. The construction process of the rigid inclusion and thus the stress generated due to it has not been modeled in the present analysis.

Table 7. Properties of the material

Properties	RI
Modulus of elasticity (MPa)	9320
Poisson's ratio (l)	0.2
Poids Volumique $\gamma\left(\frac{KN}{m^3}\right)$	24

The first calculation phase is the initial phase with K0 procedure with the original soil layers (Plaxis 2014). The second phase is the column installation with activation of the column material and of the interfaces. After this phase, the displacements are reset to zero. The last steps are the loading steps in which the maintained load are activated (Figs. 4 and 5). The loads applied to the ground are taken as a uniformly distributed load on the surface. The following diagrams show the settlements of each structure.

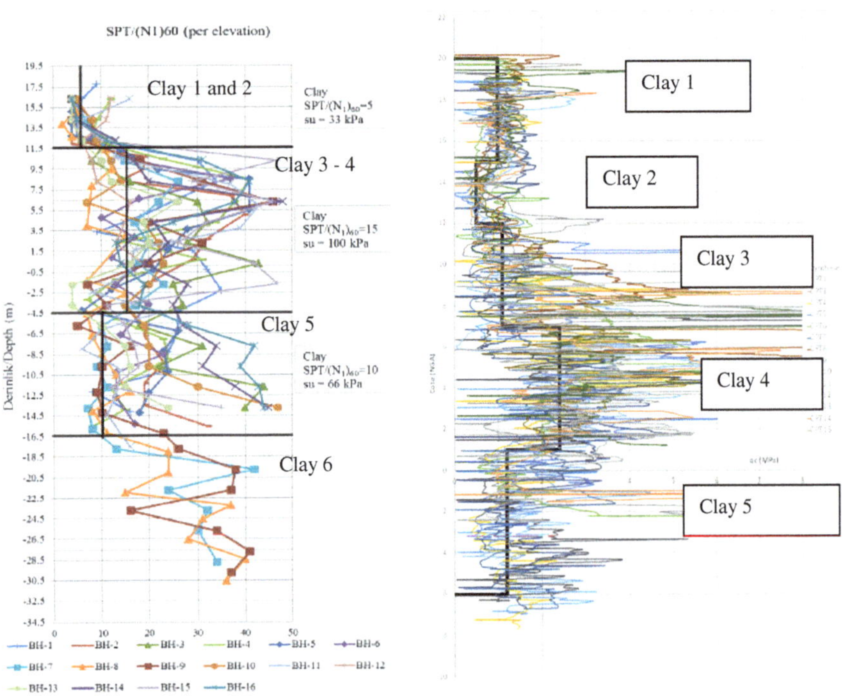

Fig. 4. Result of tests "SPT"

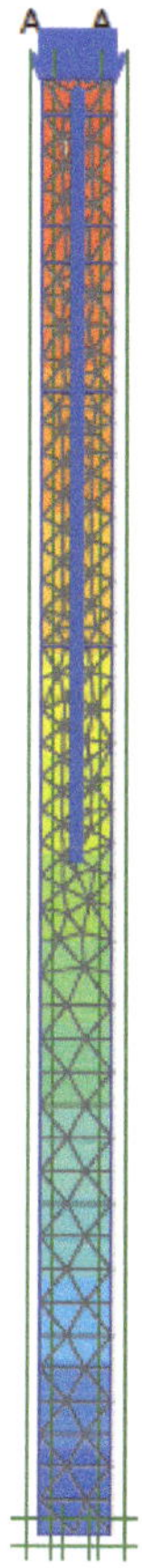

Fig. 5. Settlement of rigid inclusions (structures number 5, 6, 7)

7 Topographic Settlement Monitoring Results

Topographic settlement monitoring was conducted, several benchmarks in each structure were monitored and an average of settlements was given in the center of each structure (Fig. 6).

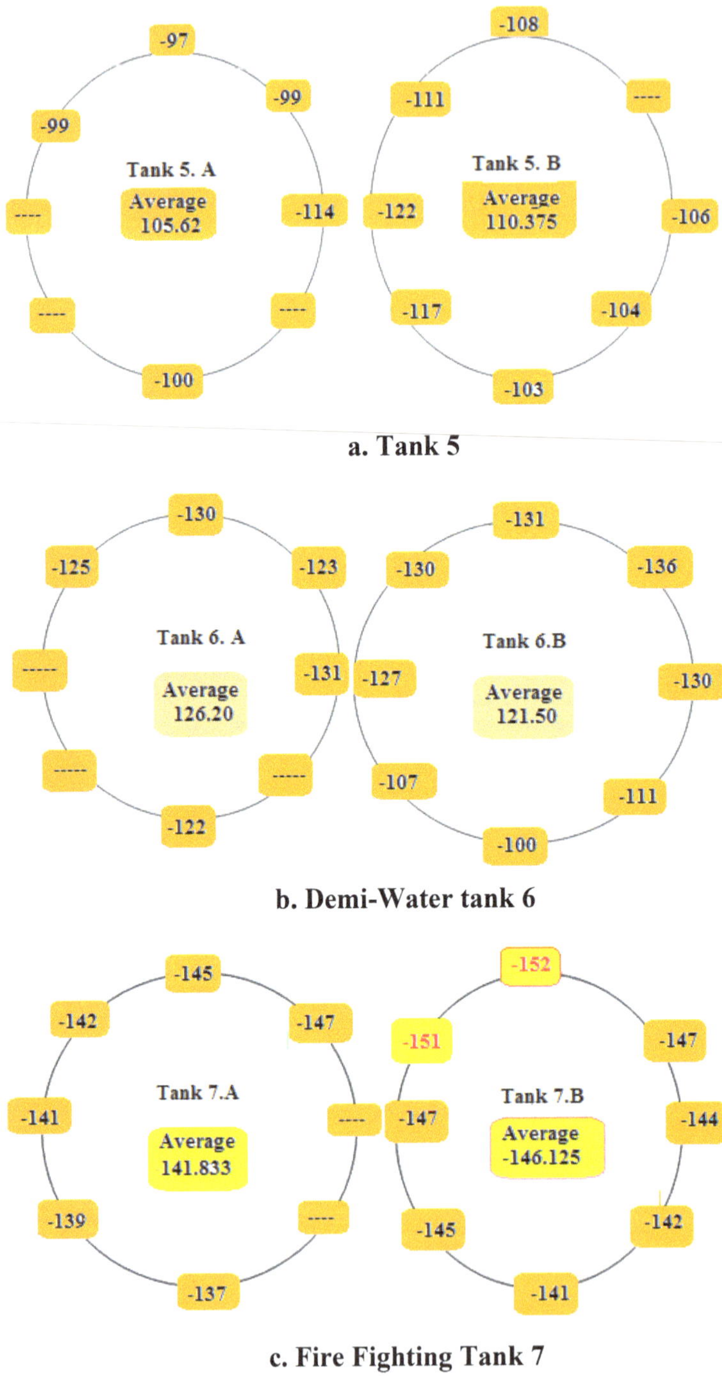

Fig. 6. Topographic settlement in mm for the structures number 5, 6 and 7

8 Comparison of Settlement Results and in Situ Measurements

The Table 8, presents the average of the final settlements of each structure and the settlements calculated by each method:

Table 8. Settlement by different methods

Structures number		Average settlement (cm)	Oedometric settlement (cm)	Numerical settlement 2D (cm)
Tank 5	A	10.5	10.5	13.3
	B	11.0		
Tank 6	A	12.6	14.5	14.3
	B	12.1		
Tank 7	A	14.18	14.2	15.3
	B	14.6		

We did not take into account the differential settlement in analytical and numerical computation, in view of the uniform distribution of the surface charge and the homogeneity of the characteristics of the soil layers horizontally as well as the modeling hypothesis of a tank unique without influence of other tank nearby.

9 Conclusions

Rigid inclusions are required to take up the majority of the load exerted by the structure. For these inclusions, there are two zones. A first zone, in the upper part of the inclusion where settlements of the latter are lower than those of the surrounding soil. We thus see a neutral point appear where the settlements of the soil and the inclusion are identical. This has been identified as being about 1/4 or 1/5 of the length of the inclusion. If we refer to the theory of rigid inclusions, we will have negative friction at the top and positive friction at the bottom. The negative friction will load the inclusion while the positive friction will have the effect of unloading it. This is what we see on the cuts made on TASPIE + results. There is a zone of maximum stress on either side of the neutral point. Inclusion is discharged by strongly positive and peak.

The comparison of the theoretical and in situ results for the works concerned shows that the differences on settlements are very limited. Some overshoot has been noted, this is necessarily due to the very large width of the structure, which does not make the choice of soil characteristics easy because the software necessarily uses homogeneous ground characteristics horizontally.

We observe the same observation for Tanks A and B of the rigid inclusion structures where there are certain settlement differences between the two tanks

The results obtained by the analytical method, even if they present certain discrepancies with the real case, seem to us very acceptable. The software uses a large number of calculation parameters (γ, Eoed, OCR, Em, qs … etc.) which reflects a perfect knowledge of the behavior of the soil, and subsequently obtaining a very satisfactory result.

The results determined by the two-dimensional numerical approach are considered acceptable with a slight overestimation of the settlement at times of the order of 2 cm and 2.8 cm for the structure 5A.

Note that the 2D modeling of a rigid inclusion network is accompanied by errors, in fact an inclusion network in view of these dimensions can not be modeled directly in 2D, the networks are considered as walls of equivalent rigidities. Such approximations are accompanied by errors, especially in the estimation of interstitial pressures.

Acknowledgements. My thanks to all the engineers at the Boufarik power plant project "GAMA, KELLER and GEOMAG POWER SYSTEMS".

References

Boussetta, S., Bouassida, M., Zouabi, M.: Assessment of observed behavior of soil reinforced by rigid inclusions. Innov. Infrastruct. Solut. **1**, 27 (2016). https://doi.org/10.1007/s41062-016-0027-6

Briançon, L., Dias, D., Simon, C.: Monitoring and numerical investigation of a rigid inclusions–reinforced industrial building. Revue canadienne de géotechnique **52**(10), 1592–1604 (2015). https://doi.org/10.1139/cgj-2014-0262

Briançon, L.: Renforcement des sols par inclusions rigides - état de l'art. Irex, paris, 185 p. (2002)

Briançon, L., Kastner, R., Simon, B., Dias, D.: état des connaissances – amélioration des sols par inclusions rigides. In: Proceedings of Symposium international sur l'amélioration des sols en place. asep-gi, paris, pp. 15–44 (2004)

Combarieu, O., Gestin, F., Pioline, M.: Remblais sur sols améliorés par inclusions rigides/premier chantier. Bulletin de liaison des laboratoires des ponts et chaussées **191**, 55–61 (1994)

Geomag: rapport de sol pour la réalisation de la centrale électrique de Boufarik (2014)

Algérie, K.: résultats des mesures de tassement sur ouvrages bâtit sur colonnes rigides du projet de la centrale électrique du Boufarik, Algérie (2017)

Glandy, M., Frossard, A.: Justification d'une fondation superficielle sur un sol renforcé d'inclusions. Annales de l'ITBTP, février (2002)

Jenck, O.: le renforcement des sols compressibles par inclusions rigides verticales. Modélisation physique et numérique. thèse de doctorat de l'insa de lyon (2005)

Simon, B.: une méthode intégrée pour dimensionner les réseaux d'inclusions rigides en déformation. In: Publication Committee of the XV ICSMGE ed. Proceedings of the 15th International Conference on soil mechanics and geotechnical Engineering, 27–31 août 2001, Istanbul, vol. 2, p. 1007. Amsterdam, Balkema (2001)

Rouag, A.: «étude critique et comparative sur le renforcement des sols par inclusions souples et rigides», master, faculté des sciences de l'ingénieur, département de génie civil, université Saad Dahleb de Blida (2017)

Vidalie, J.F.: «relations entre les propriétés physico-chimiques et les caractéristiques mécaniques des sols compressibles», rapport de recherche LPC n°65, janvier 1977

A Comparison of Soil Improvement Techniques for Frost Heave Mitigation

Yao Zhang[1(✉)], Liangxiao Xiong[2], and Lin Liu[3]

[1] Terracon Consultants, Louisville, KY, USA
zhangyao1982@gmail.com
[2] Chengdu University of Technology, Sichuan, China
xionglx1982@126.com
[3] Michigan State University, East Lansing, MI, USA
liulin6@msu.edu

Abstract. Frost action including frost heaving and thaw weakening is a common phenomenon in cold regions, which cover geographically more than half of the land area of the earth. Frost action has long been a nightmare in the management of built infrastructure in cold regions. It causes damage to transportation infrastructures and many other earth structures subjected to freezing and thawing. Cracked pavements, jacked up bridge foundations, tilted structures, malfunctioning utilities, broken pipelines, etc., all are examples of damage suffered from frost action. Billions of dollars are spent every year for repair and maintenance of the infrastructure in cold regions. Based on the records of the national transportation research group TRIP (The Road Information Program), the U.S. government spends over 2 billion dollars per year on rebuilding the thousands of miles of pavement that are destroyed by frost action, without counting the cost of the maintenance and minor damages reparations.

Soil improvement techniques for frost heave mitigation has been started ever since 1960's. The mechanism for each technique is to prevent at least one out of the three requirements for frost heave (1) frost-susceptible soil, (2) supply of unfrozen water, and (3) appropriate freezing temperature, from fulfilling. Several sustainability treatments to the soils are proved to be effective, for instance Portland cement, lime, electro-osmosis dewatering and hydrophobic layer etc. In this paper, the mechanism for each frost heave mitigation techniques are reviewed and summarized. A comparison of mechanism, suited soil types and the effects on the treated soils between different techniques are listed. The pater will be beneficial for departments dealing with infrastructures built in cold regions.

1 Introduction

In seasonal freezing regions, where the temperature cyclically drops below the freezing point of water, frost action, including frost heave and thaw settlement, frequently occurs. Frost action causes significant damage to transportation infrastructure and structures with footings in frost-susceptible soils.

When the temperature of the ground drops below freezing, the ground surface may heave. This phenomenon, called "frost heave," is caused primarily by the migration of

H. Shehata and H. Poulos (Eds.): GeoMEast 2018, SUCI, pp. 193–203, 2019.
https://doi.org/10.1007/978-3-030-01917-4_16

the unfrozen water to the frozen fringe and the formation of ice lenses in this region. Differential heave of the ground can cause bumps in asphalt pavement and cracks or movement in the concrete pavement or structural footings. When the temperature increases and the ice starts melting, the soil mass usually cannot immediately reabsorb all the resulting water (Andersland and Ladanyi 2003). The result of adding this water is an increased moisture content and loss of strength, leading to subsidence of the ground surface. This phenomenon is called "thaw weakening". Potholes in asphalt pavement and pumping in concrete pavement are examples of damage caused by thaw weakening (Zhang and Michalowski 2015).

Researchers and engineers have been trying to develop soil and site improvement techniques to mitigate this phenomenon and the resulting damage. The purpose of this paper is to review the various techniques used to mitigate frost action and compare the mechanism and effectiveness of these techniques.

2 Mechanism of Frost Heave

The primary cause of frost heave is "ice lensing," rather than the volumetric expansion water undergoes when turning into ice. The expansion of water during the phase change from liquid water to solid ice plays only a secondary role. Frost heave can cause as much as 1.5 m of displacement in some regions of the U.S. which is far greater than the volume growth that the phase change of water could cause. Additionally, early experiments in the late 1920's demonstrated that frost heave was observed even in soils saturated with fluids that contract upon freezing (e.g. benzene) (Taber 1929).

When the temperature of the soil drops below the freezing point of water, ice develops in the soil pores. The formation of ice creates an "ice lens" in the soil immediately behind the freezing front. Free water from surrounding voids and the unfrozen region of the soil becomes source of the growing ice lens. The force that drives the water from the surrounding area is called "cryogenic suction" which is caused by capillary and temperature gradient. The ice lens grows until the source of water is cut off. As the freezing front propagates through the soil, many ice lenses are generated. These ice lenses separate the soil particles and push the upper layers of soil upwards. Figure 1 shows the formation of an ice lens in a frost heave test specimen. Figure 2 illustrates the typical behavior of the soil and water near the fringe of the frozen region.

Depending on the speed of the propagation of the freezing front and the permeability of the surrounding soil, an ice lens can be formed periodically. Not all water in the proximity of the freezing front contributes to the formation of an ice lens; in fact, much of it remains in the pores and turns into a pore ice. This is often referred to as "freezing in situ" which does not produce frost heave. Ice lensing is a process that is very sensitive to both the duration and the intensity of freezing. If the temperature drops very quickly, leading to a fast propagation of the freezing front, water does not have enough time to migrate to the frozen fringe before in situ freezing occurs and large ice lenses are not formed. On the other hand, a gradual temperature change provides sufficient time for the water to migrate and supports the growth of ice lenses.

Fig. 1. Frost heave specimen with ice lenses (Lay 2005)

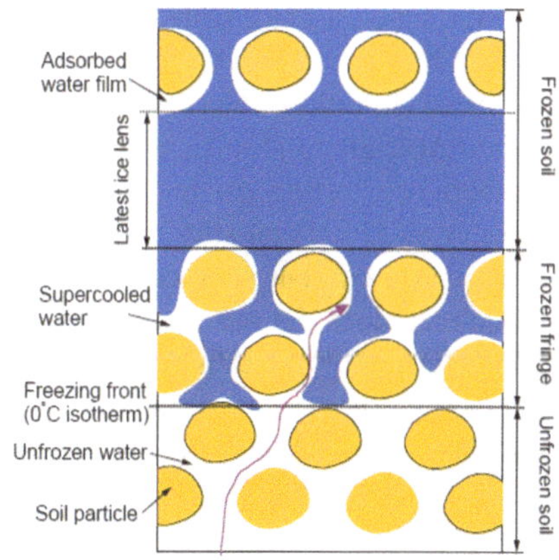

Fig. 2. Illustration of the frozen region in soil (Zhu 2006)

Soils that are prone to form ice lenses are termed frost-susceptible soils. Such soils are usually fine, but still with a good path for the migration of water during the freezing process. Frost-susceptible soils can be identified by the soil composition or by the average rate of heave observed in laboratory tests. Soils containing 3% or more grains finer than 0.02 mm by weight are typically frost-susceptible. Silts can be very susceptible to frost heaving because silt soil mixture has small voids but relatively high hydraulic conductivity, facilitating water migration and ice lensing. Clays can be frost-susceptible too, but ice lens growth in clays is slow due to their low hydraulic conductivity.

In summary, frost heave will not occur unless all the following requirements are satisfied:

(1) Frost-susceptible soil,
(2) Supply of unfrozen water, and
(3) Temperature changes that cause a slow freezing front propagation.

In the sub-sections below, methods and techniques that can be used to mitigate frost action are discussed.

2.1 Cement Treatment

The Portland cement, when sufficiently added, can bind the soil and aggregate particles together. Two stages of reaction within soil cement mixture were found, in which the first stage came out with comprise coarse platelets, and the second stage formed calcium silicate hydrate (CSH) gel. An increase in tensile strength of the matrix and a reduction in hydraulic conductivity of the treated soils can be achieved. Such an effect may increase the resistance of ice lens formation as well as narrow the path for the water migration.

However, to make this method qualified to be in the category of mitigating frost heave, the amount of cement added to a given soil should not be selected arbitrarily. The optimum amount of cement should be considered based on different soil properties and service conditions, since insufficient cement results in poor durability and excessive cement leads to shrinkage cracking.

Laboratory testing on reduced cement contents treated soil was conducted by Guthrie et al. (2007) on silty materials. Figure 3 shows the frost heave as well as the final moisture content of specimens. The results indicate a fact that frost heave of the specimen does not necessarily go down as the cement being added in. In fact, the performance is amount sensitive: insufficient amount of cement would not mitigate the frost heave of the treated soil at all. On the contrary, it would result in a greater frost heave than that occurring in untreated samples. Corresponding to the research, a cement of 2% added in the soil caused a higher frost heave and a larger amount of water uptake to the frozen zone. However, adding 3.5% and 5% of cement to the soil can almost eliminate the frost heave and the final water content after thawing were also less than the soil treated with 2% cement and no cement (Guthrie et al. 2007). It is suggested that 3 to 5% of cement is typically a good amount for silty soils, but a set of specific lab testing are recommended to optimize the parentage for an effective treatment result.

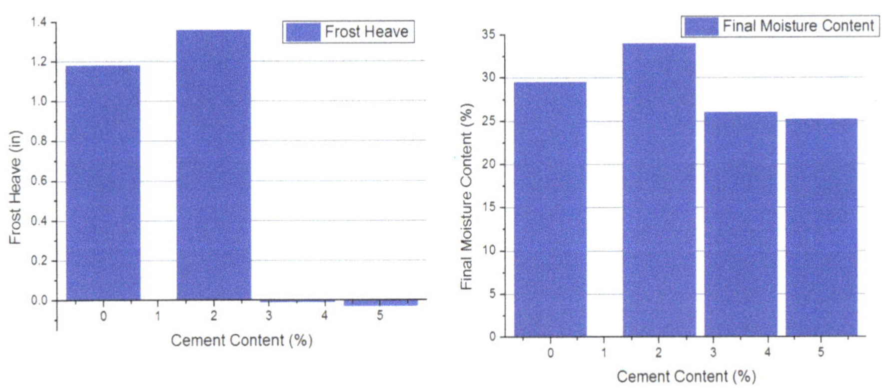

Fig. 3. Frost heave and final moisture content with different cement content (After Guthrie et al. 2007)

2.2 Lime Treatment

Researchers also studied on the availability of the treatment using lime on soil for its frost heave mitigation. A similar amount sensitivity as cement treated soil was observed. The two stages of reaction within soil lime mixture were found as well with comprise coarse platelets and CSH gel formed.

An increase in compressive and tensile strength was found to be directly related to the formation of gel, which can bind the soil particles together (Arabi and Wild 1986). In the meantime, the formation and growth of the gel tends to alter the range of the pores size, leading to a change in the permeability of the mixture. The direction of change in permeability highly depends on the amount of lime added in the mixture and the temperature under which the mixture cured. A small quantity of lime (2%) could lead to a sharp increase in permeability due to the coarsening of the pore structure and the opening of capillary channels as a result of flocculation or soil modification. Low curing temperature could also result in an increase in permeability because of the incomplete amount of gel formation and the gradual coarsening of the pore structure resulted. However, high lime contents under higher curing temperature leads to a decrease in permeability and increase in porosity. This is caused by the formation of gel, which shift the pore size distribution toward the finer pore sizes and block the path and channels in between particles (Wild et al. 1987).

Arabi et al. (1989) did serous tests on the effects of different amounts of lime (2%, 6%, 10%, 14%), different curing temperatures (25 °C, 50 °C, 75 °C) and different curing time on frost heave mitigation. The soil used in the tests is silty clay, and the mineralogical components were quartz, feldspar, and illite with traces of chlorite and hematite. Tests indicated that high curing temperature and long curing period would result in an improved performance of frost heave mitigation under same lime content. Under the same environmental conditions, higher lime content (above 2%) provided better performance on frost heave resistance. The authors suggested in the end to use more than 6% lime content to treat the frost susceptible soils in summer so that it could reach its full performance after long curing under relatively higher ambient temperatures to resist frost heave in winter.

2.3 Electro-osmosis Dewatering

Vershinin et al. (1949) got higher conductance on tested frozen ground than dry soil and pure ice, which suggested a current flow in continuous films of unfrozen water. Therefore, the ions can migrate continuously through soil matrix under electrical gradient. Hoekstra and Chamberlain (1964) then first proved by experiment the mobility of unfrozen water with the ions under electrical gradient within frozen soil. Under the electro-osmosis, the unfrozen water in frozen soil transports towards the cathode. Not even that, the ice in the region would be melting after the unfrozen water film in anode regions being depleted. The final state came out to be all the ice is removed in anode region and large ice bodies formed around cathode region. In the meantime, shrinkage cracks were observed at anode region.

The method for frost heave mitigation by electro-osmotic dewatering and meanwhile the chemical treatment (Fig. 4) was investigated by Baker and Berg (1983). Instead of just applying the electrical gradient on the frozen soil, the authors also used both flocculants (ferric chloride, $FeCl_3$) and dispersants (*Tetra-sodium pyro-phosphate* $Na_4P_2O_7$ or TSPP) to see how they could affect the electro-osmosis process. Both laboratory tests and field tests were performed to see the outcome of the frost heave mitigation as well as the mechanism behind it. For the laboratory test, soils were initially dried and then given desired chemical concentration solution to reach a

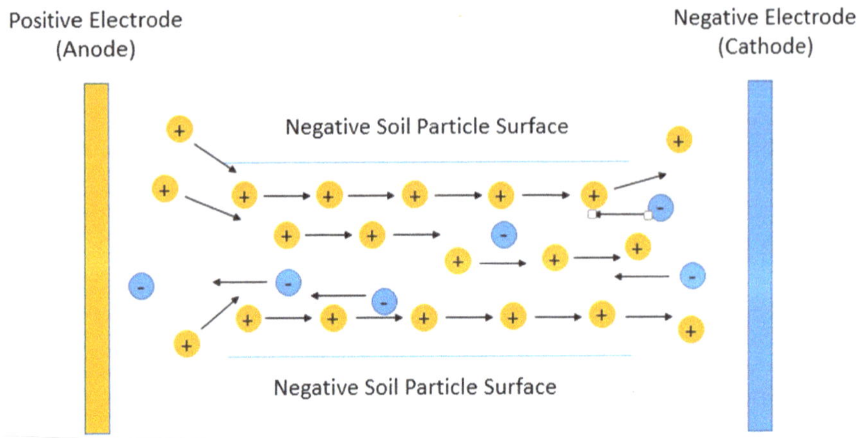

Fig. 4. Electro-osmosis flow under electrical gradient (After Baker and Berg 1983)

moisture content of 43%. After that, the soil slurries were mixed and soaked for 18 h before the direct current (DC) potential was applied to perform electro osmosis. The field tests were also placed with a larger outside diameter to achieve essentially the same design as the laboratory tests.

The test results revealed that the method of electro-osmotic dewatering with chemical treatment is excellent in reducing frost heave. In addition, a low concentration of either a flocculent or a dispersant can not only accelerate the electro-osmotic dewatering process, but reduce the power consumption significantly as well. However, high concentration of dispersant should be avoided before electro-osmotic dewatering completion due to a considerable resultant decrease in permeability which can bring down the effect of dewatering.

2.4 Hydrophobic Layer

Gieselman et al. (2008) conducted experimental study to use a hydrophobic soil layer in between the freezing front and the water table to see how this soil layer can reduce frost heave. The test results indicate that in soil cells with hydrophobic layers, the freezing front advanced more rapidly so that the surround water does not have enough time to migrate to the frozen fringe. As a result, the ice lens formation during freezing process as well as the resulted water content after thawing are less than those soils without hydrophobic layer. Because of this, the layer can greatly decrease water uptake from the water table in the winter time thus reduce the overall amount of frost heave.

2.5 Structural Polymer Foam Injection

Edgar et al. (2013) conducted a study of using structural polymer injection below pavement mitigate frost heave. The structural polymer injection creates a thermal barrier that reduces heat loss below the pavement so that the freezing front does not propagate to the subgrade.

The study was on a project site mitigating damage caused by frost heaves on the Wyoming Highway 70 in the city of Encampment, WY. A three inch (75 mm) thick layer of structural expanding polymer (Uretek 486 STAR) was injected immediately below the sub-base course at a depth of 18 in. (450 mm). The foam both supports the road surface and creates a thermal blanket to reduce heat loss. The thermal regime was determined by installing thermistors above and below the foam layer. The data shows that the temperatures below the polymer stayed at or above freezing, hence controlling the frost heave.

Surface surveying data showed that the heave was reduced from 3 to 0.5 in. (75 to 13 mm) over two winters, whereas the untreated areas still showed three inches of movement.

2.6 Capillary Barriers

Geosynthetic Capillary Barriers

Henry and Holtz (2001) investigated the potential for geosynthetic capillary barriers to reduce frost heave in soils. Soil specimens with horizontal disks of geosynthetics were subjected to freezing temperature from the top. Free water supply was made available from the base of the specimen to contribute the potential ice lens growth in the freezing process.

The research revealed that geocomposites containing dry geosynthetic and soil fines can successfully perform as capillary barriers and prevent frost heave. Moist geotextiles containing soil fines however, will not behave as capillary barriers. It is because that geotextiles have considerable wetting–drying hysteresis, thus moistened geotextiles absorb and transmit water more easily than dry geotextiles subjected to the same suction. It was also indicated that effective capillary barriers typically promote rapid frost penetration than native soils. Even though the soil below them may heave, the soil above them will have much lower water contents and thus retain strength during thaw. Consideration should therefore be given for depth of placement in a pavement section or embankment.

Recycled Gypsum and Waste Plastic Trays

Ahmed et al. (2011) investigated recycled waste gypsum plasterboard and plastic trays as an alternative for ground improvement and their potential use for frost heave mitigation. This investigation specifically examined the capillary rise time of water in soils mixed with recycled gypsum and trays. A longer capillary rise time represents the water reaching a lower height or taking a longer time to reach a given height. This means that it will take longer for the water to reach the freezing front or possibly not reach the freezing front at all in the winter time.

The researchers observed increases in the capillary rise time as recycled gypsum was added to the soil samples (Fig. 5). The increase in length of capillary rise time was attributed to the cohesion of the recycled gypsum to the soil particles. This cohesion changes the hemi-hydrate calcium sulphate in the gypsum to hydrate calcium sulphate, which is no longer reactive with water. As the gypsum content of the soil mixture increased, the length of capillary time increased. Sample age was also examined and it

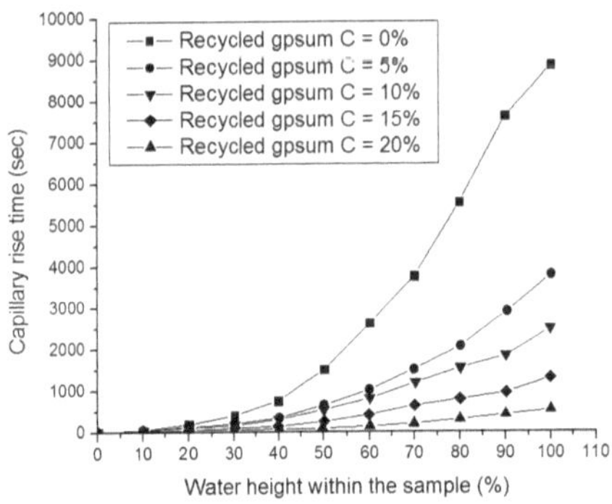

Fig. 5. Capillary rise time vs. percent gypsum content. (After Ahmed et al. 2011)

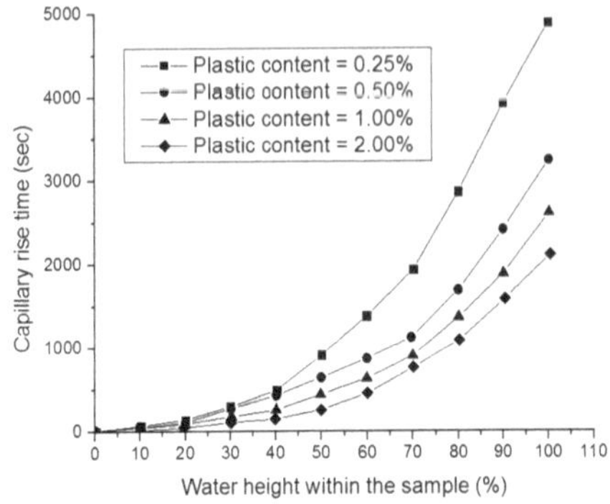

Fig. 6. Capillary rise time vs. percent plastic content. (After Ahmed et al. 2011)

was observed that younger samples had longer capillary rise times, but the times did not decrease for samples with ages beyond 14 days. The addition of plastic strips further increases the length of capillary rise time because they act as obstacles for the rise of water in the soil matrix. Also, the plastic strips are not reactive with water. An increase in the length of capillary rise time was observed as the amount and size of the plastic strips in the soil mixture were increased. Overall, a combination of 10% recycled gypsum and 2% (by mass) can potentially lengthen the capillary rise time tenfold (Fig. 6).

Soil Bags

Soil bags are comprised of coarse gravel in polypropylene bags. Soil bags have been used for many years during floods to prevent soil erosions. Recently, soil bags have been studied in Japan (Matsuoka and Sihong 2003) as a method for improving bearing capacity for foundations. Also, the potential for frost heave mitigation by using soil bags has been investigated (Suzuki et al. 2000).

The coarse gravel used in soil bags naturally has large voids. Due to the large voids, there is less capillary action in and around the soil bags reducing the level to which the water rises during the freezing process. Instead of constructing foundations to a depth past the expected frost level, surface foundations can be placed on layers of soil bags that extend past this frost depth. Suzuki et al. (2000) observed that the water content only reached 5% at a 20-cm depth for a soil layer incorporating soil bags, but reached 20% at a 20-cm depth in a soil layer with no soil bags.

Shredded Tires

Shredded tires have also been used in soil mixtures to reduce the amount of frost heave for soils in cold climates. The tires act as insulation for the soil and prevent the propagation of the wave front to greater depths.

A study of soil temperature with depth for soils with different percentages of tire shreds (by volume) was conducted by Lawrence et al. (1999). A difference of 3 °C was obtained by using a 260-mm layer of 100% tire shreds. This difference became smaller

Table 1. Comparison of different techniques

Treatment technique	Soil type	Process	Product formed	Mechanism improved[a]	Effect on soils
Lime	Clay	Chemical	Calcium Silicate Hydrate (CSH) Gel	(1)	Increase compressive and tensile strength; Reduce hydraulic conductivity
Cement	Clay	Chemical	Calcium Silicate Hydrate (CSH) Gel	(1)	Increase compressive and tensile strength; Reduce hydraulic conductivity
Electro-Osmosis	Silt & Clay	Physical	-	Immediate, but Temporary	Transport unfrozen water from anode to cathode; Melt ice lenses that
Hydrophobic Layer	Silt & Clay	Physical	-	(2)	Reduce height of rising water level; Advance frost front rapidly
Recycled Gypsum and Plastic	Clay	Chemical	Hydrate Calcium Sulphate	(1)	Gypsum reacts with soil to produce particles unreactive with water; Plastic unreactive with water and acts as obstacle for rising water
Structural Polymer Foam Injection	Silt & Clay	Physical	-	(3)	Serve as thermal barrier to reduce frost depth
Geosynthetic capillary barriers	Silt & Clay	Physical	-	(2)	Cut off the path for water to migrate to frozen zone
Soil Bags	Silt & Clay	Physical	-	(1)	Coarse gravel in bags increase the void ratio lowering capillary action
Shredded Tires	Silt & Clay	Physical	-	(3)	Insulate the soil and increase soil temperature below tire layer

[a]Each treatment tends to change the conditions of the soil that are prone to frost heave so that either (1) frost-susceptible soil, (2) supply of unfrozen water, (3) freezing temperature are improved and reduce frost heave.

as a lower percentage of tire shreds was used in the soil mixture, but was still noticeable for a 67% to 33% ratio (by volume) of tire shreds to soil. Lawrence et al. (1999) also observed a 60% decrease in the average soil heave when a 100% tire shred layer was used. The 67%/33% ratio also decreased the soil heave, but only by about 25%. Their results for the soil heave test are shown in

3 Comparison and Conclusions

All the soil treatment techniques for frost heave mitigation are summarized in the following Table 1 to compare their individual treatment mechanisms and effects on soils. Table 1 listed the feature and effects for each technique, engineers can find out and select the most suitable way for frost heave mitigation in specific engineering practice. Besides the techniques mentioned in this paper, a well-designed application of insulation will always be good method for alleviation purposes, in which condition (3) is altered.

Acknowledgments. Here you should only mention those people who actually helped in finalizing the work. Don't use this for promotion.

References

Ahmed, A., Ugai, K., Kamei, T.: Investigation of recycled gypsum in conjunction with waste plastic trays for ground improvement. Constr. Build. Mater. **25**, 208–217 (2011)

Andersland, O.B., Ladanyi, B.: Frozen Ground Engineering, pp. 102–119. Wiley, Hoboken (2003)

Arabi, M., Wild, S.: Microstructural development in cured soil-lime composite. J. Mater. Sci. **21**, 497–503 (1986)

Arabi, M., Wild, S., Rowlands, G.O.: Frost resistance of lime-stabilized clay soil. In: Transportation Research Record 1219. TRB, National Research Council, Washington, pp. 93–111 (1989)

Baker, G.C., Berg, J.C.: Reducing frost heave by electro-osmosis dewatering and soil chemical treatment. North. Eng. **15**(4), 10–16 (1983)

Edgar, T.V., Mathis, R., McGary, T., Potter, J.C.: Frost heave mitigation using structural polymer injection. In: Proceeding of Transportation Research Board 93rd Annual Meeting (2013)

Gieselman, H., Heitman, J.L., Horton, R.: Effect of a hydrophobic layer on the upward movement of water under surface-freezing conditions. Soil Sci. **173**(5), 297–306 (2008)

Guthrie, W.S., Lay, R.D., Birdsall, A.J.: Effect of reduced cement contents on frost heave of silty soil: Laboratory testing and numerical modeling. In: Transportation Research Board 86th Annual Meeting, Paper No. 07-2999, Washington, 21–25 January 2007

Hoekstra, P., Chamberlain, E.: Electro-osmosis in frozen soil. Nature **203**, 1406–1407 (1964)

Henry, K., Holtz, R.: Geocomposite capillary barriers to reduce frost heave in soils. Can. Geotech. J. **38**(4), 678–694 (2001)

Lay, R.D.: Development of a frost heave test apparatus. MSci thesis, Department of Civil and Environmental Engineering, Brigham Young University, Provo, Utah, 75 p (2005)

Lawrence, B., Humphrey, D., Chen, L.-H.: Field trial of tire shreds as insulation for paved roads. In: Proceedings of the Tenth International Conference on Cold Regions Engineering. Lincoln, New Hampshire, pp. 428–439, 16–19 August 1999

Matsuoka, H., Sihong, L.: New earth reinforcement method by soilbags ("donow"). Soils Found. **43**(6), 173–188 (2003)

Suzuki, T., Yamashita, S., Matsuoka, H., Yamaguchi, K.: Effect of wrapped gravel on prevention of frost heaving. In: The 35th Japan National Conference on Geotechnical Engineering, Japan, pp. 609–610 (2000)

Taber, S.: Frost heaving. J. Geol. **37**, 428–461 (1929)

Vershinin, P.V., Deriagin, B.V., Kirilenko, N.V.: Izvestiia Akademii Nauk SSR. Seria Geograficheskaia & Geofizicheskaia **13**(2), 108 (1949)

Wild, S., Arabi, M., Rowlands, G.O.: Relation between pore size distribution, permeability and cementitious gel formation in cured clay-lime systems. Mater. Sci. Technol. **3**, 1005–1011 (1987)

Zhu, M.: Modeling and simulation of frost heave in frost-susceptible soils. Ph.D. thesis, Department of Civil and Environmental Engineering, University of Michigan, Ann Arbor, Mich., 232 p (2006)

Zhang, Y., Michalowski, R.L.: Thermal-hydro-mechanical analysis of frost heave and thaw settlement. J. Geotech. Geoenviron. Eng. **141**(7), 04015027 (2015)

Influence of Stone Column Installation on Performance of Reinforced Foundations

Abdeldjalil Remadna(✉) and Sadok Benmebarek

MN2I2S Laboratory, Department of Civil Engineering and Hydraulic, Biskra University, Biskra, Algeria
djalilremadna@gmail.com

Abstract. This paper describes the results of numerical analysis studying the influence of stone column installation on the pressure capacity of reinforced raft with an infinite grid of stone columns. In the first part of this paper, cylindrical cavity expansion was used to evaluate the installation effects of stone column. A numerical model was created with PLAXIS 2D in axis-symmetric geometry to reproduce the lateral expansion. The results showed that the installation of a single column has a main effect on the soil stresses. These effects were represented by the increase of the effective horizontal stress and the improvement of soil stiffness. In the second part of this paper, a comparative analysis is performed on a unit cell model to investigate the influence of each installation effects on the improvement of axial pressure capacity of the unit cell. The results show that consideration of these effects leads to a significant improvement in the axial pressure capacity of the treated ground. Nevertheless, the influence of each installation effect remains dependent on the loading stage.

Keywords: Stone column · Installation effects · Numerical modeling
Lateral earth pressure · Stiffness · Axial pressure

1 Introduction

Stone columns are usually used to improve the load-settlement characteristics of soft soils. Column installation process is accompanied by vibrations and horizontal displacement on the surrounding soil which it causes further compaction of the surrounding soil. However, design of stone columns does not usually consider these installation effects and is usually based on their performance as wished in place. Field measurements have shown some of the effects of column installation (Watts et al. 2000; Kirsch 2006; Castro 2007…). Watts et al. (2000) observed the change of horizontal stress of soil surrounding the column due to the installation. The authors measured a high increase in the horizontal earth pressure (up to 60 kN/m^2) at a distance of 1.5 times the column diameter. Kirsch (2006) analyzed the changes made to stresses and pore pressure. The author shows that the installation process generates large pore pressure in the soil surrounding the column which is dissipated rapidly after the completion of the installation of the column. In addition, the lateral earth pressure coefficient (K) of the in situ soil increases from an at rest value (K_0) to a post-installation (K) value.

H. Shehata and H. Poulos (Eds.): GeoMEast 2018, SUCI, pp. 204–212, 2019.
https://doi.org/10.1007/978-3-030-01917-4_17

Numerical modeling is a useful tool that may help to derive some conclusions or recommendations about installation effects for column design (Castro and Karstunem 2010). Guetif et al. (2007) carried out a numerical procedure to simulate column installation and to estimate the improvement of Young's modulus of soft clay in which the column is considered to behave as a cylindrical cavity subjected to lateral expansion. Elshazly et al. (2008) performed a numerical analysis calibrated by load–settlement records from a full-scale field load test carried out by (Mitchell and Huber 1985) to back-calculate the post-installation lateral earth pressure coefficient value (*K*). Castro et al. (2014) are the only who reported that the installation of the column changes not only the radial stress, the vertical and circumferential stresses changes and also the values for state parameters of the soil such as the void ratio.

All the recent published numerical studies (Castro et al. 2014; Sexton and McCabe 2015; Ellouze et al. 2016) affirm that column installation can be modeled by cylindrical cavity expansion using 2D axis-symmetrical modelling. Nevertheless, these studies did not suggest the same conclusion regarding the effect of column installation. In this paper, numerical analyses were carried out to investigate stone column installation effects in a soft soil. First, cylindrical cavity expansion was modeled by Plaxis 2D in axis-symmetric geometry to evaluate the installation effects. Second, the improvement of stresses deduced from the first stage is incorporated into the second analysis to investigate the influence of installation effects on the improvement of axial pressure capacity of the treated ground.

2 Numerical Modelling

This paper deals with the numerical study of the influence of installation effects on the pressure capacity of an improved raft with an infinite grid of stone columns. Plaxis 2D code in axis-symmetric model was used to model a single column using the unit cell concept, i.e. only one column and the corresponding surrounding soil. This modelling strategy was adopted in several numerical studies to study the behaviour of an infinite grid of columns (Balaam and Booker 1981; Ambily and Gandhi 2007; Sexton et al. 2013). Since the problem is axis-symmetric, only a half of the unit cell was modelled. Figure 1 shows the unit cell model used in the analysis. Roller boundaries were assumed on the sides while the base is fixed in all directions. The displacement control method was used for the calculation of axial pressure capacity of the unit cell. The analysis was performed with differenta replacement ratio (A/A_c) which is defined by the following Eq. (1):

$$\frac{A}{A_c} = k\left(\frac{s}{D_C}\right)^2 \tag{1}$$

Where s is the spacing between columns, D_C. is the column diameter and k is a constant depend on the grid pattern (for a triangular pattern k = $(2\sqrt{3})/\pi$). The column diameter (D_C) has been fixed at 1.1 m, while the spacing between columns (s) has been varied.

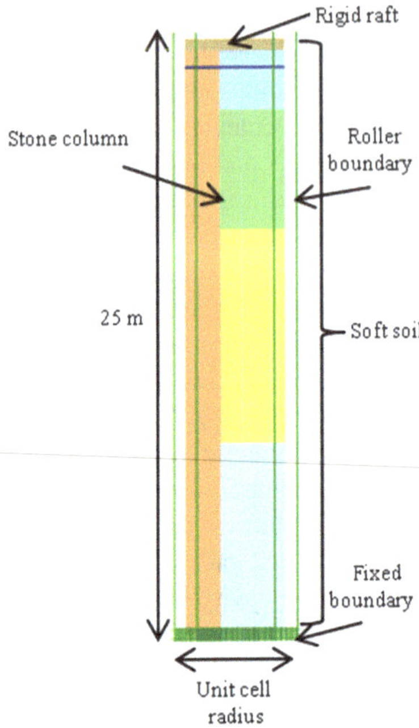

Fig. 1. Finite element model of the unit cell

The installation of the column was modelled by applying a prescribed displacement in undrained conditions. Then a consolidation analysis is performed to allow excess pore pressures to dissipate. The "up-dated" option of Plaxis software was activated because lateral expansion generates large strains. The use of this option produces higher strains and thus the mesh will be much distorted and boundary conditions may not be appropriate. Hence the estimation of the pressure capacity from this model is not suitable. Therefore, the installation effects were evaluated separately from a different finite element model. After that the new stress state (the installation effects) deduced were incorporated as an input parameter in the unit cell model.

Figure 2 shows the finite element model used to evaluate these effects. The installation of only one column was considered using an axis-symmetric model of 25 high and 30 m wide. The lateral limit considered sufficient not to influence the displacements of the soil. Roller boundaries were assumed on all sides to allow the expansion of the column.

To consider realistic situation, the profile of soil was adopted from a case study presented by Carvajal et al. (2013). Group of 13 stone columns installed in a soft soil with dry bottom feed method up to a depth of 25 m. The soft soil consists of 4 layers with silty sand and clayey silt up to 25 m. The water table is located at 1.2 m below the surface. The material properties of the stone column were chosen from the literature since it was not presented by Carvajal et al. (2013). These values fall between the typical ranges of actual problems. Both stone column and soft soil were modelled by Mohr-coulomb behaviour using drained material properties. The parameter values for 4

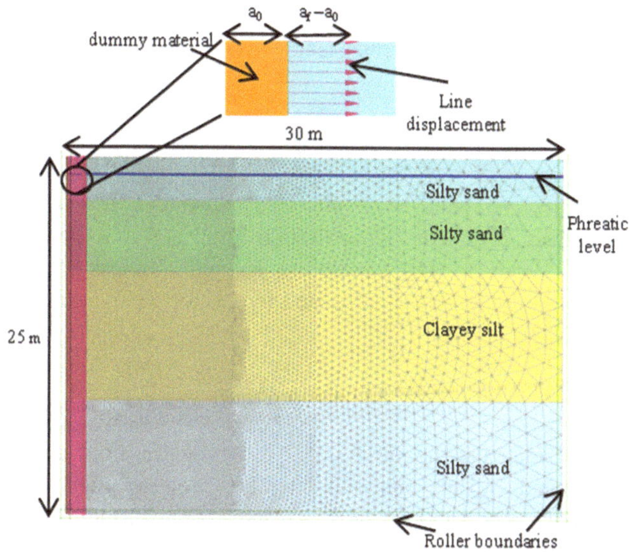

Fig. 2. Finite element model for the column installation

Table 1. Soil properties

Layer	Depth (m)	γ_{sat} (kN/m^3)	c' (kPa)	ϕ' (°)	E_{oed} (kPa)	k_h (m/s)
Silty sand	0–3	18.5	1	33	20000	$1.15 * 10^{-8}$
Silty sand	3–8	18.5	5	25	4500	$1.15 * 10^{-8}$
Clayey silt	8–17	18.5	10	17	3600	$2.30 * 10^{-9}$
Silty sand	17–25	18.5	5	28	7200	$1.15 * 10^{-8}$
Stone column	0–25	19	1	45	70000	$1.97 * 10^{-5}$

layers of soil and stone column are listed in Table 1. The rigid raft of 0.5 m thickness was modelled as a linear elastic material with $\gamma = 24$ kN/m^3, $E = 3.10^7$ kPa and $\nu = 0.15$. These values represent the properties of ordinary concrete typically used in numerical studies.

3 Results

3.1 Evaluation of the Installation Effects

The results showed that the installation of the column has a main effect on the soil stresses after the consolidation. The radial displacement of the soil due to the column installation leads to a significant increase in the effective radial stress (σ'_r), and consequently an increase in the effective mean stress (p') of the soft soil. Therefore, the change made to the radial stress is represented by an improved lateral earth pressure coefficients (K) defined as:

$$K = \sigma'_r / \sigma'_{z0} \quad (2)$$

Where σ'_r is the effective radial stress after the consolidation and σ'_{z0} is the initial effective vertical stress. Moreover, Kirsch (2006) indicated that the increase of mean effective stress (p') leads to an increase in the soil stiffness (E). Therefore, the improvement of soil stiffness (E) is calculated by the following relation (Biarez et al. 1998; Guetif et al. 2007):

$$E/E_0 = p'/p'_0 \quad (3)$$

Where the index "$_0$" indicates the initial state. Consequently, stone column installation effects are represented by an improved lateral earth esre coefficient (K) and an improved stiffness (E). The variation of K/K_0 and E/E_0 with the radial distance (r) is plotted in Figs. 3 and 4 respectively.

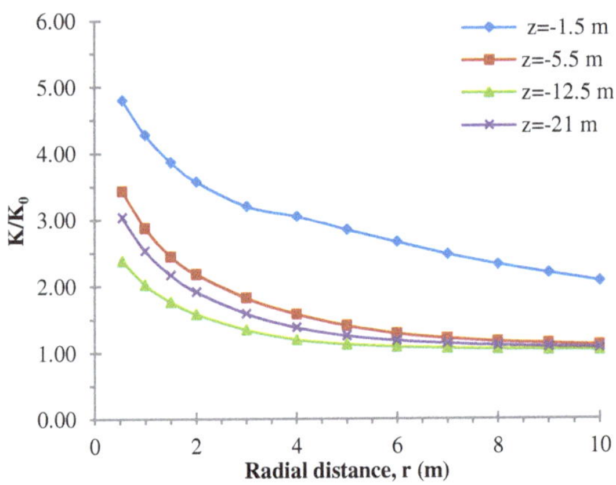

Fig. 3. Variation of K/K_0 with radial distance (r)

3.2 Axial Pressure Improvement

The results were presented on a term of the axial pressure improvement ($API = q_{tr}/q_{untr}$) which defined as the ratio of axial pressure on the unit cell of the treated ground with stone column (q_{tr}) on that of untreated ground (q_{untr}) at the same applied displacement (U_y). The API was evaluated for different cases of the installation effects:

- Case A: no installation effect (K_0 and E_0);
- Case B: installation effect represented by the improvement of K;

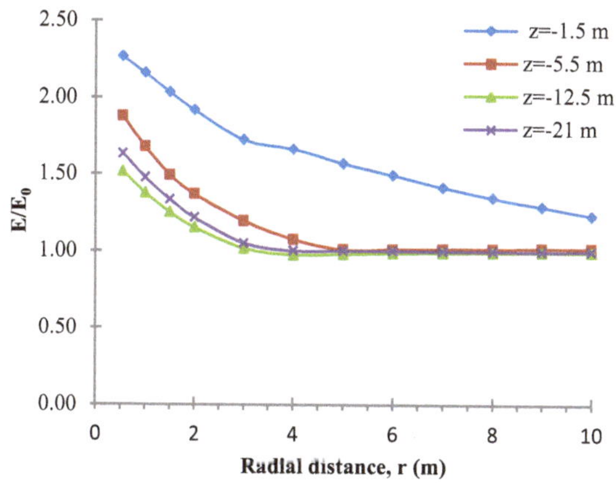

Fig. 4. Variation of E/E_0 with radial distance (r)

- Case C: installation effect represented by the improvement of Young's modulus of soil E;
- Case D: global installation effect represented by improving both effects K and E;

Figure 5 shows the variation of *API* versus area replacement ratio (A/A_c) for the different analyzed cases, and under a displacement of 0.2 m (10% of A). It is clear that the *API* values decreases as the A/A_c increases with and without installation effects. A similar result was predicted by Ambily and Gandhi (2007) on experimental and FEM analysis. The authors reported that as the column spacing increases the axial pressure decreases. In Fig. 5, it is obviously seen that for any value of A/A_c the consideration of installation effects (B, C and D) predict a higher value of *API* than the case with no installation effect (case A). Also, the highest value of *API* was predicted when the global installation effect (cases D) was considered. The comparison between cases B and C shows that there is a negligible difference between the *API* values for any value of A/A_c. This means that each effect has a positive influence on the improvement of axial pressure capacity. However, it appears that the most significant improvement should be achieved with the consideration of both effects (lateral earth pressure coefficient and soil stiffness).

Figure 6 shows the variation of *API* versus imposed displacement (U_y) with and without installation effects for $A/A_c = 3.6$. The values of *API* for the cases with installation effects (B, C and D) are obviously higher than that with no installation effect. In the same line of thought of cases A and C, the improvement of axial pressure would be decreased if (U_y) increases. In contrast, the *API* in the cases B and D still unchanged up to a displacement of $U_y = 0.2$ m. Beyond this, the improvement slightly decreases. Regarding the comparison between each installation effect, Fig. 6 shows that the *API* values is higher in the case C than case B at small displacement ($U_y < 0.2$ m), while it is not at larger displacement ($U_y > 0.2$ m). At low displacement, *API* is mainly influenced by Young's modulus (E) value and the effect of increasing K has a little influence on the axial pressure. This is due that, at low displacement, the soil has been in an elastic behaviour and strains have changed only as a function of Young's

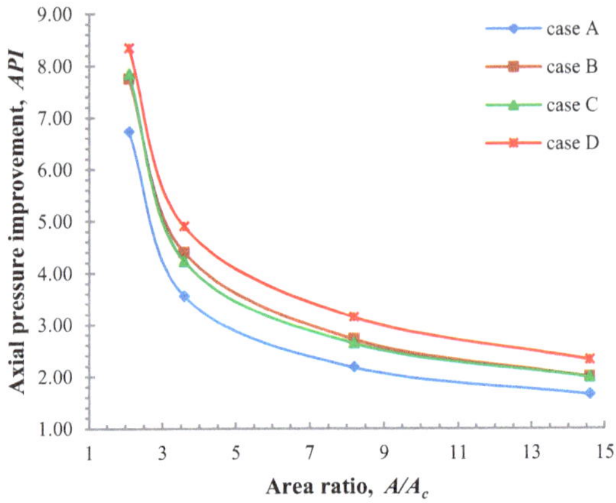

Fig. 5. Variation of the axial pressure improvement (*API*) with A/A_c under a displacement of 0.2 m

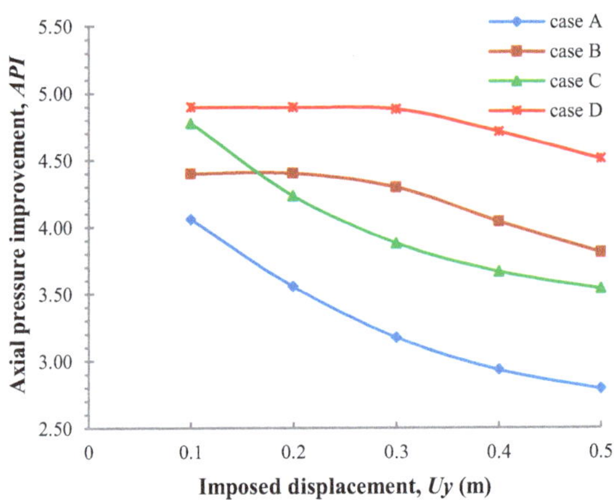

Fig. 6. Variation of the axial pressure improvement (*API*) with imposed displacement (U_y) for A/A_c= 3.6

modulus (*E*) value (according to Hooke's law). In contrast, at larger displacement, the improvement of axial pressure is more influenced by the values of lateral earth pressure coefficient (*K*), whereas the effect of Young's modulus (*E*) has been negligible. This is due that at high displacement the soil has been in a plastic behaviour and strains were not controlled only by Young's modulus (*E*) value (Hooke's law is not applicable here). This confirmed by the plot of plastic points in Fig. 7. It is apparent that the number of plastic points is higher in cases A and C compared to cases B and D. This is due that in the case A and case C the column was yielded so the improvement was decreased. On the contrary, the soil in case B and D has higher values of the lateral earth pressure coefficient (*K*). Larger *K* and increases the lateral confinement which gives more improvement to the axial pressure capacity of the treated ground.

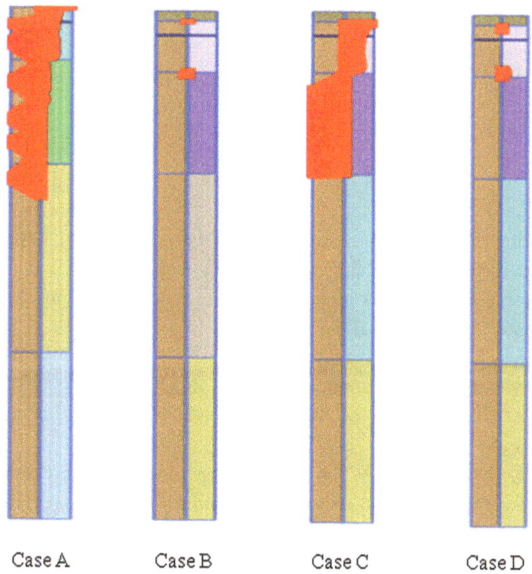

Fig. 7. Spread of plastic point in the unit cell

4 Conclusions

Numerical investigations were carried out to study the influence of installation effects on the axial pressure of an infinite grid of columns using the unit cell concept. The installation of a single stone column was modelled as a cylindrical cavity subjected to lateral expansion. Stone column installation effects are represented by two major effects:

- The increase of lateral earth pressure coefficient (*K)* due to the improvement of effective radial stress;
- The increase of soil stiffness (*E*) due to the improvement of effective mean stress (p').

The installation effects were considered as input parameters to represent the improvement of soil due to the column installation. The comparison between each installation effect showed that:

- The effect of increasing the *K* reduces the spread of plasticity in the unit cell and gives more confinement to the stone column. However, this effect has a negligible influence at lower displacement.
- The effect of increasing stiffness modulus (*E*) reduces the elastic strains and increases the strength of the unit cell. However, this effect has a significant influence only at low displacement.
- The highest improvement is predicted when the installation effects are represented by both effects (lateral earth pressure coefficient and stiffness modulus).

Finally, installation effects should be considered in the numerical studies of stone column ground improvement. The installation effects must be expressed by the improvement of stiffness modulus (*E*) and the lateral earth pressure coefficients *K*.

References

Ambily, A.P., Gandhi, S.R.: Behavior of stone columns based on experimental and FEM analysis. J. Geotech. Geoenviron. Eng. **133**(4), 405–415 (2007)

Balaam, N.P., Booker, J.R.: Analysis of rigid rafts supported by granular piles. Int. J. Numer. Anal. Meth. Geomech. **5**(4), 379–403 (1981)

Biarez, J., Gambin, M., Gomes-Corriea, A., Falvigny, E., Branque, D.: Using pressuremeter to obtain parameters of elastoplastic models for sands. In: Proceedings of the First International Conference on site Characterization, Atlanta Georgia, USA, 19–22 April 1998, pp. 747–52 (1998)

Carvajal, E., Vukotić, G., Castro, J., Wehr, W.: Comparison between theoretical procedures and field test results for the evaluation of installation effects of vibro-stone columns. In: Hicks, et al. (eds.) International Conference on Installation Effects in Geotechnical Engineering, pp. 205–211. Taylor and Francis Group, Rotterdam, London (2013)

Castro, J.: Pore pressures during stone column installation. In: Proceedings of the 18th European Young Geotechnical Engineers Conference, Ancona, Italy (2007)

Castro, J., Karstunen, M., Sivasithamparam, N.: Influence of stone column installation on settlement reduction. Comput. Geotech. **59**, 87–97 (2014)

Castro, J., Karstunem, M.: Numerical simulations of stone column installation. Can. Geotech. J. **47**(19), 1127–1138 (2010)

Ellouze, S., Bouassida, M., Bensalem, Z., Znaidi, M.N.: Numerical analysis of the installation effects on the behaviour of soft clay improved by stone columns. Geomech. Geoeng. Int. J. (2016). https://doi.org/10.1080/17486025.2016.1164903

Elshazly, H., Elkasabgy, M., Elleboudy, A.: Effect of inter-column spacing on soil stresses due to vibro-installed stone columns: interesting findings. Geotech. Geol. Eng. **26**, 225–236 (2008)

Guetif, Z., Bouassida, M., Debats, J.M.: Improved soft clay characteristics due to stone column installation. Comput. Geotech. **34**(2), 104–111 (2007)

Kirsch, F.: Vibro stone column installation and its effect on ground improvement. In: Triantafyllidis, T., (ed.) Proceedings of the International Conference on Numerical Modelling of Construction Processes in Geotechnical Engineering for Urban Environment, pp. 115–124. Taylor and Francis Group, London, Bochum, Germany (2006)

Mitchell, J.K., Huber, T.R.: Performance of a stone column foundation. Geotech. Eng. ASCE **11**(2), 205–223 (1985)

Sexton, B.G., McCabe, B.A., Castro, J.: Appraising stone column settlement prediction methods using finite element analyses. Acta Geotech. (2013). https://doi.org/10.1007/s11440-013-0260-5

Sexton, B.G., McCabe, B.A.: Modeling stone column installation in an elasto-viscoplastic soil. Int. J. Geotech. Eng. **9**(5), 500–512 (2015)

Watts, K.S., Johnson, D., Wood, L.A., Saadi, A.: An instrumented trial of vibro ground treatment supporting strip foundations in a variable fill. Géotechnique **50**(6), 699–708 (2000)

Utilization of Plastic Straws to Improve Lateritic Soil in Paving

Lucas Gabriel Lopes da Silva(✉), Thamara Silva Barbosa, and Ivonne Maria Alejandra Gutierrez Gongora

UCB, Universidade Católica de Brasília, Brasília, Brazil
lukkasgabriel0@gmail.com

Abstract. Knowing that compulsively discarded plastic straws are degrading the environment, it is necessary to find new, environmentally correct, and technically feasible destinations for them. The present article aims to evaluate the mechanical properties of Brasília's lateritic soil when incorporating polymeric fibers from recycled plastic straws randomly distributed in the soil. Brasília's lateritic soil and recycled plastic straws were used for this research and 0.25, 0.5, and 0.75% of them were analyzed in relation to the soil mass. The soil was collected at Universidade Católica de Brasília and was duly characterized and studied. It was necessary to characterize the soil through laboratory tests, such as liquidity limit, limits of plasticity and granulometry, and characterize the polymers of recycled plastic straws. For the mechanical characterization of the soil, the California Bearing Ratio (CBR) tests and simple compression with triplicate samples were performed for each percentage. The study was performed using the normal compaction energies (26 strokes in 3 layers of the test specimen) and modified energy (27 strokes in 5 layers of the test specimen). Thus, as indicated by the tests of the California Bearing Ratio and simple compression, it was observed that the straws increased the penetration resistance values of the CBR, most of them are higher than the penetration resistance value of the natural soil and increase the cohesion in normal energy for the simple compression test.

1 Introduction

Considering the real importance that paving has in Brazil as the basis of the main land transportation mean, it is relevant that the techniques used in paving try to solve the problems of environmental degradation that are generated by it, such as the employment of recycled materials for soil improvement in asphalt pavements.

Thus, the use of recycled plastic straws is a great advance to slow the flow of solid waste to the environment. It may even cause improvements in the mechanical properties of soils by using them as geosynthetic fibers, since the straws have polyvinyl chloride, polypropylene, and polyethylene.

In literature it is possible to find authors whose works are made with plastic fibers, which are for soil improvement. The use of random fibers in the soil-cement, according to Specht's (2000) research, improves the mechanical properties of the soil - especially the cracking control increases the post-rupture load capacity, the durability and ductility of the cemented material.

H. Shehata and H. Poulos (Eds.): GeoMEast 2018, SUCI, pp. 213–223, 2019.
https://doi.org/10.1007/978-3-030-01917-4_18

Also, in relation to the increase of mechanical strength, it is important to analyze the proper compaction energies that are used. Junior (2005) states that the CBR values boost significantly with the addition of compaction energy.

The techniques of the materials used in paving have met with more advanced technologies, and according to de Senço (2008) it can generate savings in construction of an enterprise. In light of this, it is important to analyze that the technology can be part of a system of sustainable development that guarantees the construction of enterprises without ostensively compromising the natural resources.

Assis and Melo's research (2012) examines the technical viability of soil with pellet granulated PET in pavement layers, using mixtures of soil with pellets from 0.25% to 4%. The ISC and the expansion's results point to soil improvement.

According to Santos and Silva (2015), once the efforts are distributed between the soil matrix and the fibers, the latter, that are important, shows an improvement. The authors also say that the soil matrix and the fibers are being used in modern times because they have a wide range of applicability, such as landfill reinforcement, slopes, and erosion management.

It is interesting to note that plastic waste can also work as geosynthetics, because plastic fibers generally offer high tensile strength.

According to the literature, the incorporation of PET bottle fibers increases the resistance of the soil when compared to the natural soil.

This paper aims to contribute with the analysis of the behavior of soil-fiber interaction and to verify the mechanical behavior of the soil when it incorporates plastic straws fibers with the percentages of 0.25%, 0.5% and 0.75% in relation to the soil mass, using two compaction energies (Normal and Modified).

2 Material and Methods

In order to carry out this research, the following steps were implemented: bibliographic study, soil classification, molding of test specimens in triplicate with normal and modified energy and the analysis of laboratory results using Excel software.

Studies on subjects involving the analysis of soil-fiber interaction, geosynthetics, and compaction energy variation were initially analyzed. Priority was given to studies on natural soil's mechanical behavior and fiber addition.

Subsequently, it was necessary to make characterization tests in the laboratory to characterize the type of soil collected: Liquidity limits, according to the ABNT NBR 6459: 1984; plasticity limit, according to the standard ABNT NBR 7180: 1984; and the determination of granulometric distribution, according to the standard ABNT NBR 7217: 1987. Figure 1 shows the equipment and soil for the consistency tests and Fig. 2 shows the equipment for the Casagrande characterization.

After the soil characterization study, the mechanical characterization process was initiated through compaction tests in order to identify the optimal moisture of natural soil and mixtures, according to ABNT NBR 7182: 1986, California - California support index Bearing Ratio (CBR), according to the Brazilian Standard DNIT 172: 2016; simple compression, according to the Brazilian standard ABNT NBR 12770: 1992.

Fig. 1. Materials for limits of plasticity

Fig. 2. Casagrande characterization test

Figure 3 shows the molding process of the test pieces with plastic straws, Fig. 4 shows the test specimen, Fig. 5 shows the simple compression testing mechanism, and Fig. 6 shows the ruptured test specimen.

The simple compression test was performed with a digital strain gauge to evaluate vertical deformations and two analog strain gages (one at each end) to assess the

Fig. 3. Mold molding process

Fig. 4. Proof body

horizontal deformations. The results of the latter were obtained through the average of the results of the two horizontal extensometers.

After the laboratory tests with natural soil, 0.25%, 0.5%, and 0.75% of plastic straws, an analysis of the simple compression and CBR's results were performed with the aid of Excel software.

Fig. 5. Simple compression mechanism

Fig. 6. Proof body broken in “x”

3 Test Parameters

3.1 Soil Characterization

The results of soil characteristics, such as grain size and consistency, will be presented first.

The soil's granulometry refers to well-graded soil, with predominating silt and fine sand. The grain size curve is shown in Fig. 7.

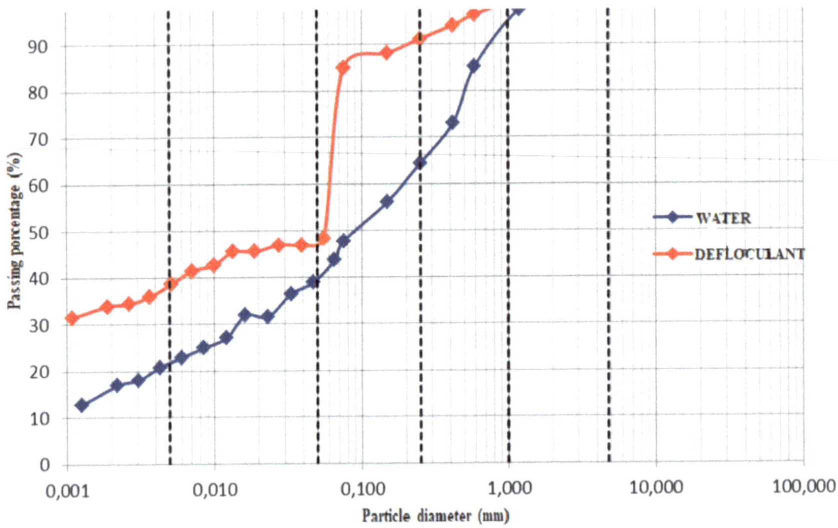

Fig. 7. Granulometry curve with and without deflocculant

The plasticity index's value was found to be 10.24% and, according to the SUCS classification, the soil is classified as clayey silt of medium plasticity which corroborates to the particle size test. Figure 8 shows the graph of the soil's liquidity limit.

3.2 Mechanical Characterization

The optimum moisture found for Brasília's lateritic soil and for the mixtures in modified and normal energy is present in the compaction curves of Figs. 9 and 10.

The CBR test to find the penetration resistance was performed in the 4 day condition. Tables 1 and 2 show the detailed results of the optimum moisture, specific dry mass, expansion, and ISC in modified and normal energy, respectively.

The simple compression test was also performed for the two compaction energies. Tables 3 and 4 show the results in modified and normal energy respectively. Figures 11 and 12 show the simple compression curves.

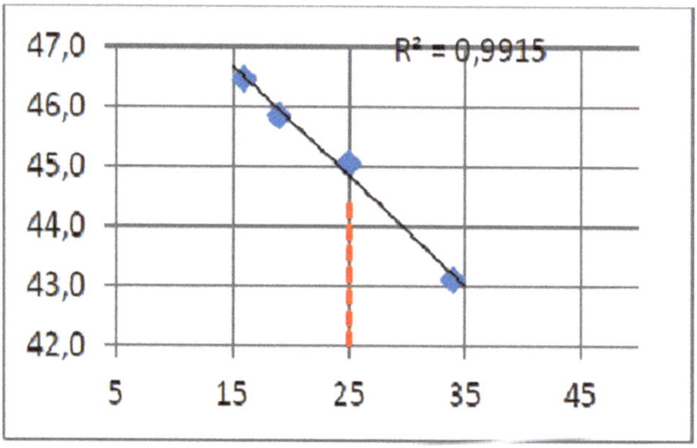

Fig. 8. Liquidity limit line

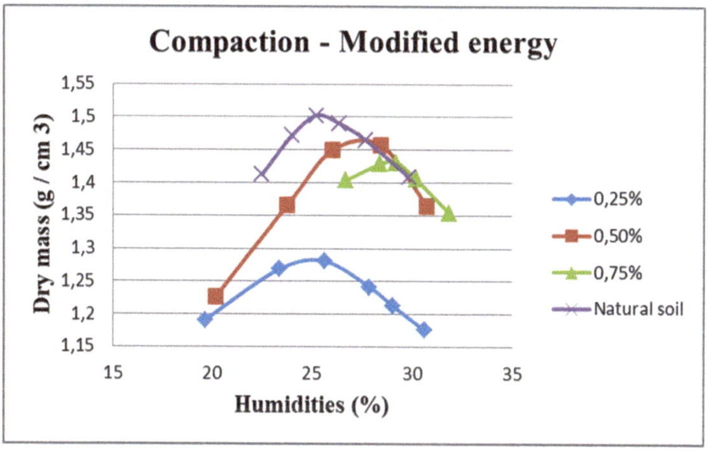

Fig. 9. Compaction curves in modified energy

4 Results and Discussions

First, for the compaction test the data found in this article is in agreement with the literature, as Assis and Melo (2012) noticed a reduction of the maximum dry mass with the addition of "Pelete". They also evaluated a range of values that grows concomitantly with the ISC, increasing its quantity to a point where the ISC falls, indicating that there is an optimal percentage of "Pelete".

The addition of plastic straws had a behavior that was similar to the previously mentions plastic fragments, because it also reduces the specific mass with the increment of plastic straws for the modified compaction energy. The percentage of 0.25% shows the fall of maximum dry mass (in relation to the natural soil), its growth with 0.5%, and another fall with 0.75% of straws. The specific mass of the soil decreases with the addition of straws in both energy, but the dry mass of 0.25% of straws for the normal

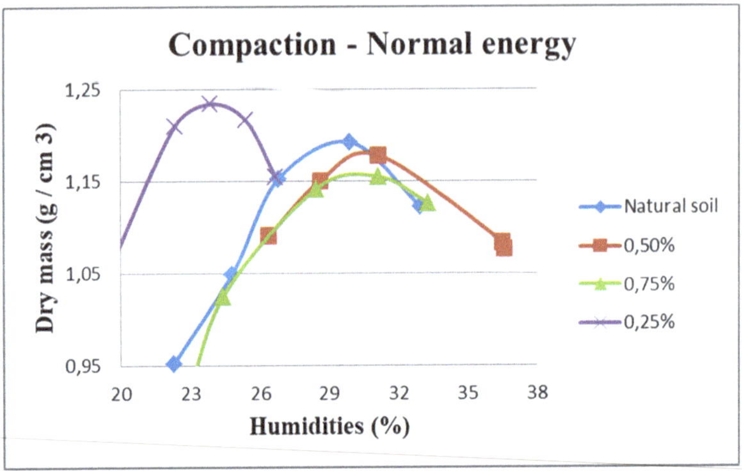

Fig. 10. Compaction curves in normal energy

Table 1. Optimum humidity, specific dry mass, expansion and penetration resistance in modified energy

Material	Optimal humidity (%)	Dry mass (g/cm^3)	Expansion (%)	Penetration resistance (%)
Soil	25,4	1,5	0,044	53,52
Soil + 0.25%	25	1,26	0,34	50,14
Soil + 0.5%	27,5	1,46	0,07	53,91
Soil + 0.75%	28,9	1,43	0,09	52,46

Table 2. Optimum humidity, specific dry mass, expansion and penetration resistance in normal energy

Material	Optimal humidity (%)	Dry mass (g/cm^3)	Expansion (%)	Penetration resistance (%)
Soil	29,6	1,19	0,044	50,04
Soil + 0.25%	25	1,23	0,035	50,91
Soil + 0.5%	30,5	1,18	0,053	51,01
Soil + 0.75%	31	1,155	0,052	50,43

energy is higher than the natural soil, 0.5% and 0.75% of straws (it may indicate an elevated resistance for the 0.25% of straws).

This study also presented a range in which the penetration resistance grows and then decreases. In modified and normal energy, the penetration resistance values had peaks in 0.5% recycled plastic straws and later the value dropped when 0.75% of plastic straws are added.

Table 3. Simple compression in modified energy

	Cohesion (KPa)	Vertical deformation (%)	Horizontal deformation (%)
Soil	252	0,20	0,52
Soil + 0.25%	272,5	0,64	1,11
Soil + 0.5%	135,5	0,31	0,52
Soil + 0.75%	115	0,31	1,4

Table 4. Simple compression in normal energy

	Cohesion (KPa)	Vertical deformation (%)	Horizontal deformation (%)
Soil	27,91	0,85	1,85
Soil + 0.25%	98,2	0,68	0,76
Soil + 0.5%	65	0,06	1,64
Soil + 0.75%	63	0,24	1,24

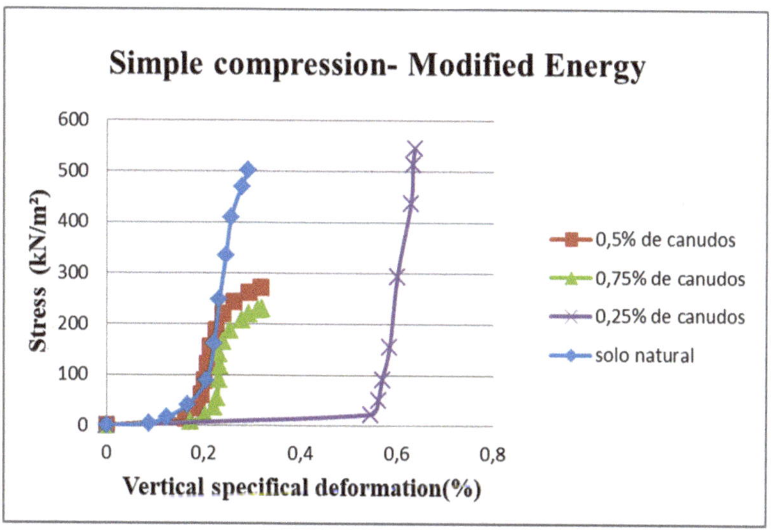

Fig. 11. Simple compression in modified energy

For the modified energy, it was observed that the penetration resistance in 4 min of assay is superior to the soil in 0.5% of straws but decreases when 0.75% of straws are added. For the normal energy, it was observed that the penetration resistance increases according to the increment of the percentage of straws but decreases in the percent of 0.75%.

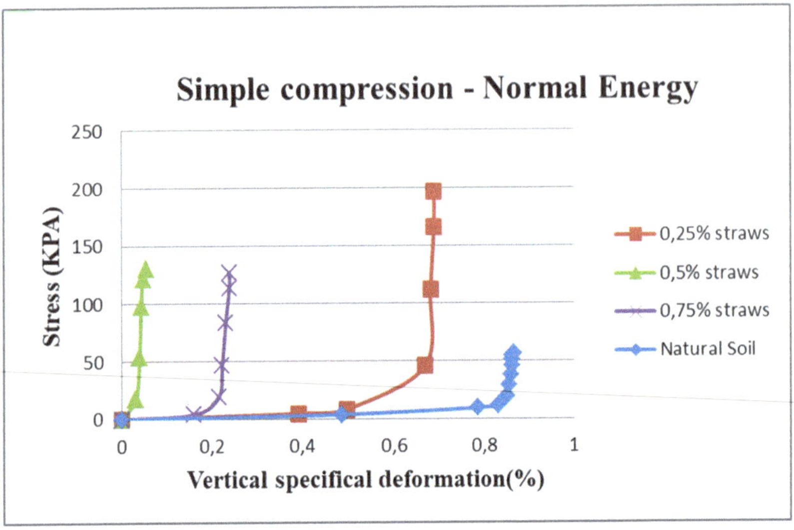

Fig. 12. Simple compression in normal energy

According to the results of CBR, it was also observed that the natural soil expansion in the two energies were equal. The results of penetration resistance in modified energy are mostly superior to penetration resistance values in normal energy.

It is also important to mention that in modified energy and normal energy, most expansion values are higher than the natural soil, a fact which can be explained by some voids that the plastic straws leave in compacted soil. However, the percentage of 0.25% of plastic straws had the highest dry density (1.23 g/cm^3), lower expansion in relation to natural soil (0.035%), and higher penetration resistance than in nature soil (50.91%), slightly below the highest penetration resistance value in normal energy in 0.5% of straws (51.01%).

It is observed that the test of simple compression done in normal energy showed better cohesion results in relation to the modified energy. The plastic straws for all percentages of normal energy were quite satisfactory, especially for 0.25% of straws that presented the best cohesion value. It is also observed that with the addition of higher percentage values of plastic straws the cohesion value drops.

For the modified energy in the simple compression test the best percentage found was 0.25% of straws, since it increased the soil support capacity, but with higher vertical and horizontal deformations, which generated a greater absorption of loads. For the normal energy the best percentage found was also 0.25% of straws, but the horizontal and vertical deformations are smaller than in the natural soil.

The cohesion values in modified energy may have been lower in the percentages of 0.5% and 0.75% in relation to the natural soil cohesion because the random distribution of straws in higher percentages may have generated weaknesses in the test specimen, similarly to the fall of the penetration resistance to 0.75% of plastics straws. However, in normal energy, all percentages of straws were superior to the cohesion of the natural soil.

5 Conclusion

It is recommended to use 0.25% of straw in normal energy, since the results of specific dry mass were superior to the soil (with higher resistance) and to all straw percentages. It presented the promising penetration resistance and was very close to the highest value found. It was also very satisfactory for the simple compression test, because it presented greater cohesion and smaller vertical and horizontal deformations in relation to the soil and smaller horizontal deformation in relation to the other percentages tested.

The use of plastic straws can be used for paving since it meets the expansion specifications (less than 2%) and the penetration resistance of DNIT (National Department of Transport Infrastructure of Brazil) in light traffic.

Acknowledgments. All the authors of this article participated in the research.

References

Assis, S.R.D.H., Melo, R.A.D.: Viabilidade técnica do uso de misturas de solo e pet granulado em camadas de pavimentos. 18ª RPU - Reunião de pavimentação urbana. São Luís, MA (2012)

Associação Brasileira de Normas Técnicas. NBR 6459: Limite de Liquidez: Referências (1984)

Associação Brasileira de Normas Técnicas. NBR 7180: Limite de Plasticidade: Referências (1984)

Associação Brasileira de Normas Técnicas. NBR 7217: Granulometria. Referências (1987)

Associação Brasileira de Normas Técnicas. NBR 7182: Compactação: Referências (1986)

Associação Brasileira de Normas Técnicas. NBR 12770: Compressão Simples: Referências (1992)

de Senço, W. Manual de técnicas de pavimentação. PINI ltda, São Paulo, SP, pp. 437–438 (2008)

Junior, J.D.S.: O efeito da energia de compactação em propriedades dos solos utilizados na pavimentação do estado do ceará. Universidade federal do Ceará, Fortaleza (2005)

Santos, J.R., Silva, T.J.L.: Análise da viabilidade de utilização de solo tropical com a adição de fibras pet em sistemas de cobertura de aterro sanitário. UNB, departamento de engenharia civil e ambiental, Brasília (2015)

Specht, L.P.: Comportamento de misturas solo-cimento-fibra submetidos a carregamentos estáticos e dinâmicos visando a pavimentação. Programa de pós-graduação em engenharia civil da universidade federal do Rio Grande do Sul, Porto alegre (2000)

A Comparative Study of the Treatment of Oil Field Produced Water by Adsorption and Membrane Intervention

Debasish Konwar, Subrata Borgohain Gogoi(✉), Joyshree Barman, and Bwhwithi Machahary

Department of Petroleum Technology, Dibrugarh University, Dibrugarh, Assam, India
subrata@dibru.ac.in

Abstract. Produced water (PW) is a complex mixture of organic and inorganic compounds and the largest volume of by-product generated during oil and gas recovery operations. Large volumes of water are produced from various oil fields of Upper Assam Basin. The potential of oilfield PW to be a source of fresh water for water-stressed oil-producing countries and the increasing environmental concerns in addition to stringent legislations on PW water discharge into the environment have made PW management a significant part of the oil and gas business. PW contains different toxic compounds that adversely affect the environment such petroleum contaminants and impurities are oil, naturally occurring radioactive minerals (NORM), oil & grease (O&G), hydrocarbons (HC), production chemicals and various metals. This paper deals with the characterization of PW generated from different depths and horizons of Upper Assam, India oil fields. PW separated from oil and gas is either injected into the reservoir for pressure maintenance or discharged to the environment. The evaluation and characterization of PW contaminants is targeted in this study. This will be done with the help of adsorption by powdered activated charcoal and membrane treatment technology by using hollow fibre membranes, which will enable for the safe disposal to the environment and reservoir pressure maintenance for improved oil recovery. The evaluation included the analysis of pH, conductivity, salinity, total dissolved solids (TDS), turbidity, O&G, hardness, alkalinity and inorganics. The results are validated in Lenntech software to determine the corrosion forming tendency of PW on the basis of Ryznar stability index and Langelier saturation index.

Keywords: Produced water · Petroleum contaminants
Adsorption of powered activated charcoal · Membrane treatment
Environment

1 Introduction

It has been observed that the oil industry is handling more water than oil and hence it can be referred as the, "water industry", [1]. In the United States, oil wells produce about 7 to 10 barrels of water for each barrel of oil [2]. It was estimated that, about 210

H. Shehata and H. Poulos (Eds.): GeoMEast 2018, SUCI, pp. 224–245, 2019.
https://doi.org/10.1007/978-3-030-01917-4_19

million bbl of water was produced daily in 1999 worldwide and is an integral and unavoidable part of oil and gas recovery processes [1, 3, 4]. This PW includes the formation water and water resulting from drilling, production and workover activities [5, 6]. Out of various options for management of PW, mostly adopted options by the hydrocarbon exploration industries is the underground injection for pressure maintenance and increased oil recovery [7, 8]. But in most cases PW is discharged to the surrounding environment [2, 5]. The quality of PW varies considerably depending on the geographic location of the reservoir oilfield, the geochemistry of the producing formation, the type of hydrocarbon being produced and the characteristics and type of the producing well [10, 11, 15].

In North-East India PWs are the by-products generated during various operations in the hydrocarbon exploration industries are major sources of pollution in the area. These pollutants find their way and traverse to various environmental receptors causing potential danger of polluting the same. The water bodies, both surface and underground are more susceptible to such pollution in oil fields. Management of PW is a key issue for the oil and gas upstream industry because of its huge generated volume and high handling cost. The aging oilfields of Upper Assam basin belonging to Oil India Limited, Duliajan and Oil and Natural Gas Limited, Nazira, produces crude oil with high water cuts even more than 90%. Production of PW from these aging oilfields is in increasing trend as oil production continues. This PW is highly saline, very rich in dissolved minerals and has a high temperature than the normal atmospheric water temperature. The salinities of the PWs from these oil fields range are relatively lower from 1000–3000 mg/l compared to other oil fields of the world [11, 15]. The oil field in this study area disposes PW in an average of 3000 m^3 on a daily basis, which is an environmental concern since it has a potential to migrate and contaminate the ground water as well as the ecosystem of the area.

1.1 Adsorption Treatment

Adsorptions of PW on powdered activated charcoal (PAC) remove organic chemicals and reduce toxicity in some PWs to allow safe discharge into surface water and is a widely accepted technology for treating and removing of organics, free chlorine, color and many other impurities. PAC can be made in particulate form as powders or fine granules less than 1.0 mm in size with an average diameter between 0.15 and 0.25 mm. Thus, they present a large surface to volume ratio with a small diffusion distance (Fig. 1). PAC is porous, inexpensive and readily available for use as adsorbents, exhibiting a large surface area to remove contaminants. It is widely used for the removal of petroleum contaminants and impurities from PW due to its high capacity for adsorption of such compound, arising from their large surface area and porosity. Figure 1 illustrates the absorption mechanism of activated carbon. Pores (macropores and micropores) on PAC use to trap different sizes of contaminants. Macropores are used to trap larger contaminants and micropores are used to trap the smaller molecules. The amount and distribution of pores will affect the efficiency of the absorption in PAC.

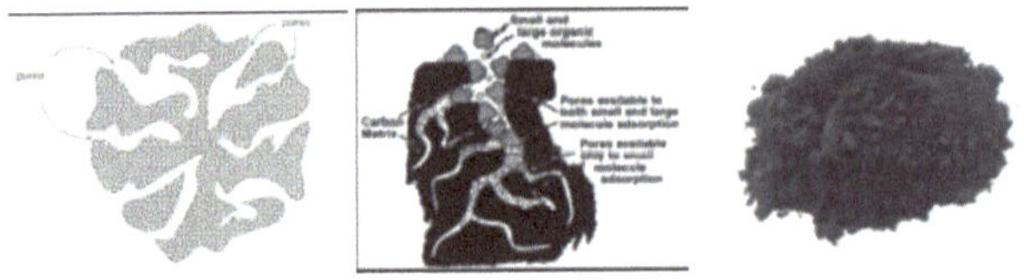

Fig. 1. Powdered activated charcoal samples

1.2 Membrane Treatment

A membrane is a barrier that separates the two phases and controls the transmission of different chemical components in a certain approach. There are generally two phases in a membrane process that are physically separated from each other by the third phase (membrane). The phases are composed of components that one of the mixture components is transferred more than the others. In other words, the membrane has selectivity to one of the components. Hence, the transition of that component from one phase into another will be carried out by the membrane. Thus, one phase of that component is enriched and the other is depleted. Such a functioning can be seen in Fig. 2 [12].

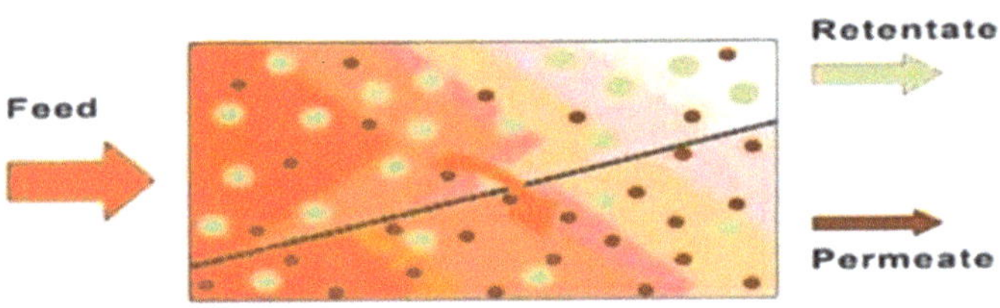

Fig. 2. Schematic model of the membrane functioning

During the treatment process of PW, the pollutants is deposited on the membrane which may lead to fouling effect that will affect the membrane's performance. It is a major obstacle to the widespread use of this technology. Membrane fouling can cause severe flux decline and affect the quality of the water produced. Severe fouling may require intense chemical cleaning or membrane replacement. This increases the operating costs of a treatment plant. There are various types of foulants: colloidal (clays, flocs), biological (bacteria, fungi), organic (oils, polyelectrolytes, humics) and scaling (mineral precipitates).

Membrane permeability (Lp) determines the membrane is porous or not and the extent to which it is porous.

Mathematically

$$Lp = \frac{J^0}{\Delta P} \tag{1}$$

Where J^0 = Steady state flux for distilled water

ΔP = Transmembrane pressure (TMP)

2 Experimental

2.1 Materials

Ten samples of PWs were collected from Oil Fields of Upper Assam. Oil samples were collected from well heads along with PW (Table 1).

Table 1. Materials used in the experiments

SN	Materials	Specification	Source
1	PWs samples	Barail formation	OIL E & P
2	EDTA (Ethylene Diamine Tetra Acetic Acid)	0.01 M	RFCL Limited
3	Erichrome black indicator	0.5 g in 100 ml	
4	Sodium Chloride (NaCl)	M.W. 58.44 g/mol	
5	Hydrochloric Acid (HCl)	M.W. 36.46 g/mol, Sp. Gr. at 25 °C is about 1.18	
6	Phenolphthalein	pH range 8.2–10 (colourless to pink)	
7	pH buffer	pH 10	WTW
8	Potassium Chloride (KCl)	M.W. 74.56 g/mol	Avantor Performance Materials India Limited
9	Calcium Carbonate ($CaCO_3$)	M.W. 100.09 g/mol	Merck Specialities Pvt. Ltd.
10	Lithium Carbonate (Li_2CO_3)	M.W. 73.89 g/mol	Spectrochem Pt. Ltd. Mumbai (India)
11	Petroleum Ether	Wt. per ml at 20 °C, 0.630–0.645 g	Fisher Scientific
12	Methyl Orange	–	Human Diagnostics & Surgichem

2.2 Instruments

The instruments that were used for the experimental work is shown in Table 2.

2.3 Methodology

2.3.1 Physico-Chemical Parameters

pH, salinity, EC, TDS, DO, turbidity were determined in Water Analyser (Table 2).

Table 2. Equipments used in the experiments

SN	Instruments	Make	Model
1.	Water Analyser	SYSTRONICS	Water Analyser 371
2.	Flame Photometer	SYSTRONICS	Flame photometer 128 Compressor 126
3.	Atomic Absorption Spectrometer	PERKIN ELMER	Atomic Absorption Spectrometer AAnalyst 200
4.	Rotospin	TARSONS	–
5.	Centrifuge	ELECTRA	Electra lab centrifuge

Table 3. Table of Ryznar stability index

Sample	Ryznar stability index (RSI)	Indication	
		Ryznar, 1942	Carrier, 1965
PW-1	8.9	Water is very aggressive	Heavy corrosion
PW-2	12		Corrosion intolerable
PW-3	13		
PW-4	10		
PW-5	11		
PW-6	11		
PW-7	11		
PW-8	12		
PW-9	12		
PW-10	11		

2.3.2 Inorganic Parameters

Na, K, Ca, Li were determined in Flame Photometer (Table 2) and Fe, Cu, Cr, Mn were determined in AAS (Table 2).

2.3.3 Validation of Results

The results were validated in Lenntech software to determine the corrosion forming tendency of PW on the basis of Ryznar stability index and Langelier saturation index (Table 2).

3 Results and Discussions

3.1 Adsorption Treatment

3.1.1 Physico-Chemical and Inorganic Characterization of PW Before and After Adsorption Treatment

Ten numbers of PW samples were collected from different wells of Upper Assam Basin and the samples were diluted in the ratio from 1:1 to 1:10 with distilled water. 100 ml of diluted samples were prepared and stored in tarsons. Prior to treatment, the samples

Table 4. Table of Langelier saturation index

Sample	Langelier Saturation Index (LSI)	Indication	
		Langelier (1936)	Improved Langelier by Carrier (1965)
PW-1	0.30	Water is supersaturated with respect to calcium carbonate ($CaCO_3$) and scale forming may occur	Slightly scale forming and corrosive
PW-2	−1.4	Water is undersaturated with respect to calcium carbonate. Undersaturated water has a tendency to remove existing calcium carbonate protective coatings in pipelines and equipment	Serious corrosion
PW-3	−1.9		
PW-4	−0.75		
PW-5	−0.75		
PW-6	−0.17		Slightly corrosive but non-scale forming
PW-7	−0.80		Serious corrosion
PW-8	−1.2		
PW-9	−0.39		Slightly corrosive but non-scale forming
PW-10	−0.71		

were first characterized and the parameters were determined. Then PAC of 0.035 g was added to each sample. These PAC added samples were kept for 2 h and mixed for 30 min using rotospin for the proper adsorption of adsorbate i.e., impurities present in the samples into adsorbent i.e., the PAC.

Later the samples were centrifuged using centrifuge for the separation of PAC from the samples and finally the samples were filtered using filter paper. Then the same parameters were again analyzed and then compared with the before treatment column in Table 5. The comparison values of the experimented values before and after treatment of PAC adsorption is shown in Table 5. These values are compared with WHO specifications for its safe disposal.

3.1.2 Treatment of Physical and Chemical Contaminants of PW

Table 5 is the comparison of the values obtained from the experimental study to the standard of World Health Organization (WHO) specification for safe disposal of water to the environment. The samples were treated in the laboratory to make within the accessible limit of WHO specification so that it can be used for agricultural purpose, re-injection to the reservoir, re-use in any industry, and safely discharge into the water bodies and environment. The physiochemical status and graph obtained from the experimental study of the PW samples are shown in Figs. 3, 4, 5, 6, 7 and 8.

In the Fig. 3, it was clearly seen that the pH of maximum samples after adsorption treatment lies between 7.5 to 9 i.e. the PW were alkaline in nature and can be reused or discharged into the river.

Table 5. Comparison of formation water quality with WHO specification after adsorption treatment

Parameters	Specification	Present study report	
	WHO	Before treatment	After treatment
pH	6.0 to 9.0	6.01–9.05	7–9.79
Cond. (mS)	0.25–0.75	0.287–27.24	0.357–21.9
TDS (ppt)	24.59	0.152–11.8	0.188–3.85
DO (ppt)	.007	0.0017–0.0045	0.0271–0.087
Salinity (ppt)	0.2–0.4	0.14–3.63	0.18–3.28
Hardness (ppt)	0.3	0.07–0.512	0.05–0.21
Fe (ppt)	0.003	0–0.006	0
Cu (ppt)	0.003	0	0
Cr (ppt)	0.002	0–0.003	0
Mn (ppt)	0.050	0–0.004	0
Na (ppt)	0.002	0.0563–0.7	0.0223–0.664
Ca (ppt)	0.075	0.00651–0.687	0.0033–0.696
O&G (ppt)	0.25	2–3.5	0
Turbidity (NTU)	5–10	0.12–117	0.06–77

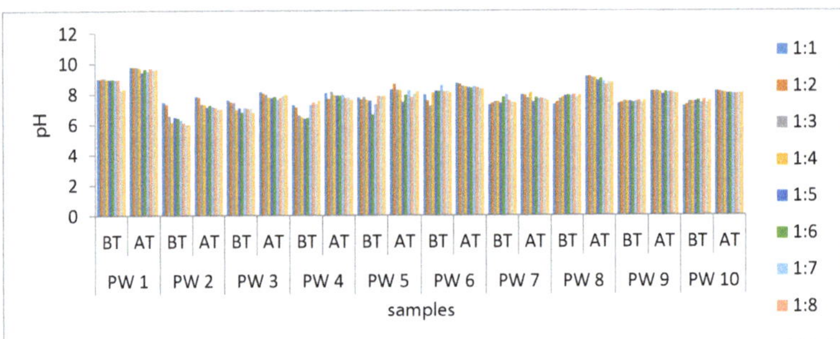

Fig. 3. Comparison diagram of pH after treatment

In the Fig. 4, the graph showing the salinity of the PW samples in which salinity reduces upto a certain value after PAC treatment except sample PW 1 and PW 8. Electrical Conductivity (EC) is a measure of the capacity of a solution to carry an Electric Current, and the permissible value of EC for water to discharge into the environment is 0.25 to 0.75 mS/cm. In Fig. 5, it is seen that EC reduces after adsorption treatment but still the PW was highly conductive. The high value (0.55 mS) of EC above pure water can be attributed to the fact that PW has ions [12]. Total Dissolved Solids (TDS) shows good result which is shown in Fig. 6. Turbidity is another test used to indicate the quality of drinking water with respect to colloidal and residual suspended matter. Increases in turbidity are often accompanied with increases in pathogen numbers, including cysts or oocyst [16]. Most of the result of turbidity of PW samples lies under the permissible limit of WHO specification as shown in Fig. 7,

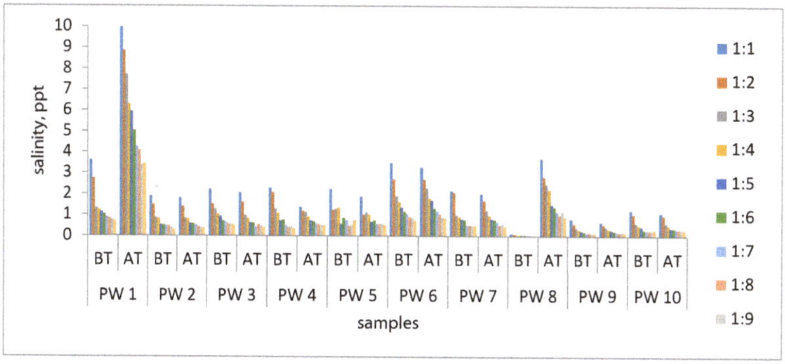

Fig. 4. Comparison diagram of Salinity after treatment

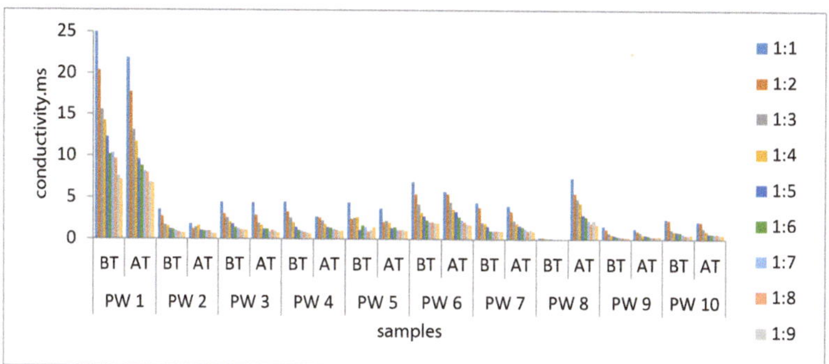

Fig. 5. Comparison diagram of conductivity after treatment

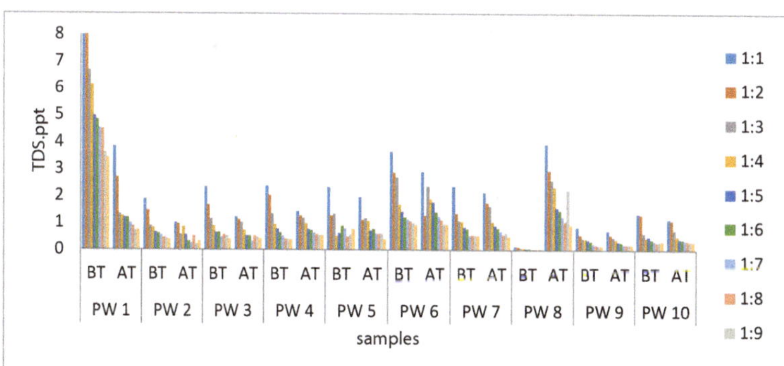

Fig. 6. Comparison diagram of TDS after treatment

which signifies that it will not affect the aquatic plant and animals and can be discharge to the river or stream. After treatment there is a huge reduction in turbidity up to approximately 70–90%. But samples 6 & 9 reports above permissible limit after treatment. The measurement of turbidity is based on comparison of the intensity of

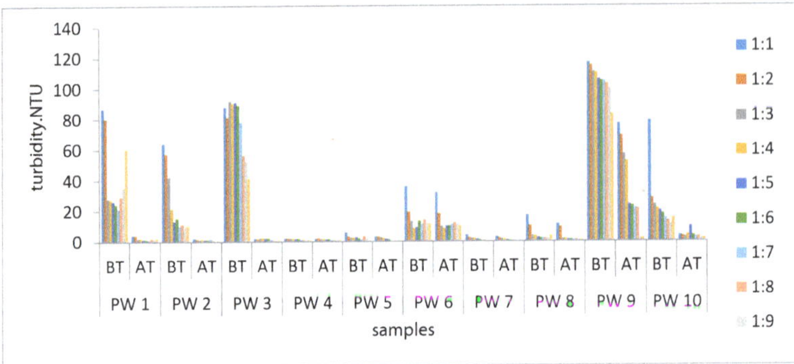

Fig. 7. Comparison diagram of turbidity after treatment

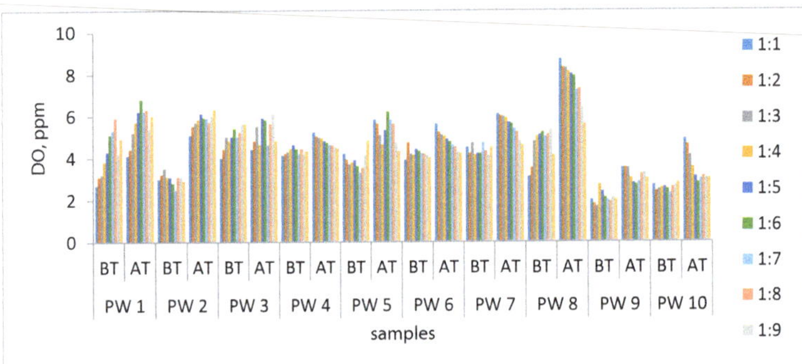

Fig. 8. Comparison diagram of DO after treatment

light scattered by a sample to the light scattered by a reference suspension under the same conditions. Colloidal matter will scatter or absorb light and thus prevent its transmission. It should be noted that the presence of air bubbles in the sample will cause erroneous turbidity readings. The sample cell or tube which was used to store the water sample should be clear, colorless glass or plastic. The cell or tube must be cleared both inside and outside, and discard if scratched or etched. Dirt or finger print may affect the result [10]. Since Dissolved Oxygen (DO) is an important water quality parameter for various purposes. DO levels in surface water body indicate the ability to support aquatic life. In the present study, DO increases after treatment process by PAC removes the synthetic organic chemicals from water as shown in the Fig. 8. Thus, adsorption of oxygen increases the concentrations of acidic oxide groups on the carbon surface, resulting in an increased base sorption capacity particularly in the ranges of the carboxylic acid group [5]. Figure 9 shows the comparison between the hardness before adsorption treatment and after treatment. It can be seen that there is much reduction of hardness in some of the samples like sample no. 4, 6, 7, 8, 9 and 10, but the rest of the samples shows slight reduction. Hardness, a physico chemical property of water, is generally a measure of calcium and magnesium ions in water [19]. The high values of hardness are detrimental for human health. O&G in PW is due to the presence of suspended oil particles in the water samples.

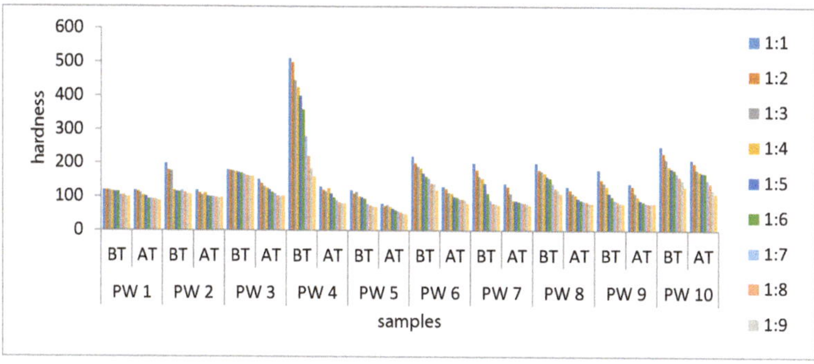

Fig. 9. Comparison diagram of hardness after treatment

3.1.3 Characterization and Treatment of Inorganics Present in PW

Figures 10, 11, 12 and 13 represents the inorganic contaminants in PW samples before and after treatment obtained from the parameter were determined by Flame Photometer and AAS. These inorganic contaminants are Na, Ca, K, Li, Cu, Cr, Fe & Mn. After experiment, high amount of Na, Ca, K, Li and some traces amount of Cr, Fe and Mn were found in the PW samples but were reduced after going through PAC filter.

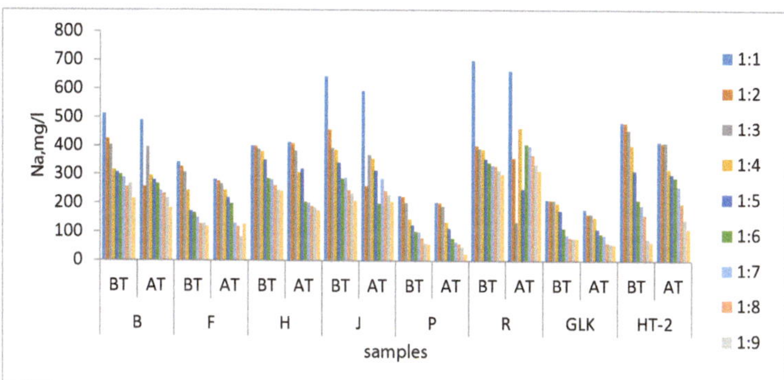

Fig. 10. Comparison diagram of Na after treatment

3.2 Membrane Treatment

3.2.1 Physico-Chemical Characterization of PW Before and After Membrane Treatment

Table 6 is the comparison tables of the values obtained from the experimental study to the WHO specification which is a standard for safe disposal of water to the environment. EC, Hardness, O&G, Na showed values much higher than the safer limits for its disposal. Whereas DO value showed lower values than the permissible limits which indicates low level of free, non-compound oxygen present in water [11].

Table 7 shows that the most of the parameters of PW showed values higher than the maximum permissible limits for safe disposal. The pHs of the ten samples were almost

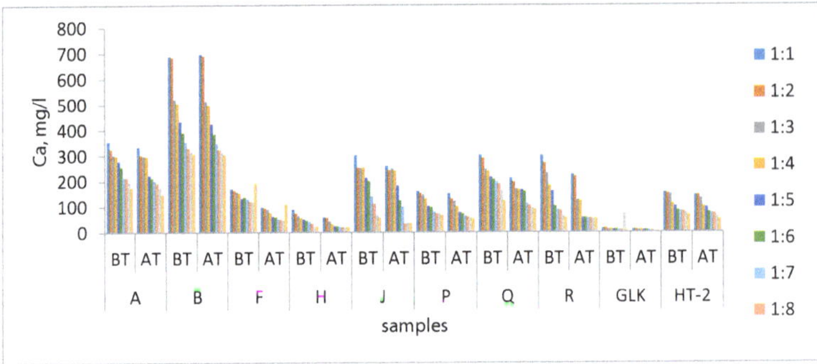

Fig. 11. Comparison diagram of Ca after treatment

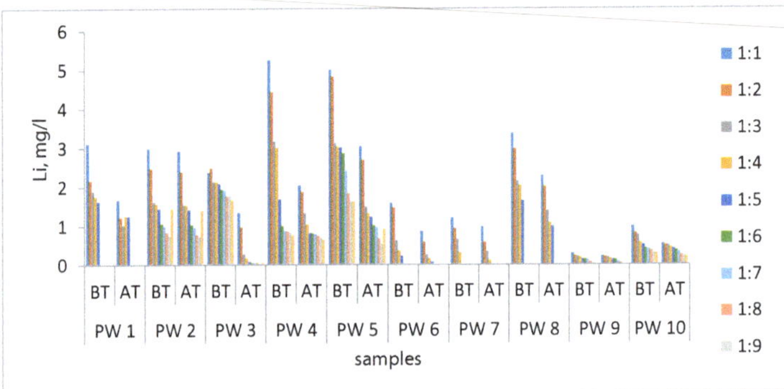

Fig. 12. Comparison diagram of Li after treatment

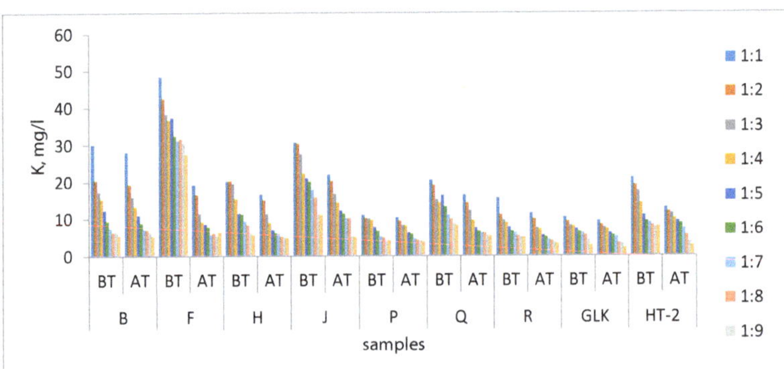

Fig. 13. Comparison diagram of K after treatment

neutral. The high value of EC can be attributed to the fact that PW has ions so the conductivity is well above pure water which is 0.55 mS [12]. PW has dissolved and suspended solids which lead to increase in TDS. DO signifies the presence of gaseous oxygen (O_2) present in water. DO values showed lower values than the discharge

Table 6. Comparison of formation water quality with WHO specification after membrane treatment

Parameters	WHO specification	Present study report	
		Before treatment	After treatment
pH	6.0 to 9.0	7.4–7.99	6.14–7.83
EC (mS)	0.25–0.75	1.43–6.5	0.0402–6.42
TDS (ppt)	24.59	0.8–3.89	0.1–3.39
DO (ppt)	0.007	0.004–0.044	0.0051–0.072
Salinity (ppt)	0.2–0.4	0.8–4.5	0.04–3.28
Turbidity (NTU)	5–10	1.5–5.3	0.09–3.1
Hardness (ppt)	0.3	0.068–0.7	0.03–1.1
O&G (ppt)	0.25	2–3.5	0
Na (ppt)	0.002	0.026–0.6132	0.0167–0.387
Ca (ppt)	0.075	0.021–0.522	0.00365–0.403

regulation values which is harmful for the aquatic animals. The high values of turbidity are due to the presence of suspended and dissolved solids in PW. Higher turbidity levels are often associated with higher levels of viruses, parasites and some bacteria because they can sometimes attach to dirt present in water [13]. Hardness, a physico-chemical property of water, is generally a measure of calcium and magnesium ions in water [14]. The high values of hardness are detrimental for human health. O&G in PW is due to the presence of suspended oil particles in the water samples.

Table 7. Physico bio-chemical characterization of OFPW

Parameters	PW-1	PW-2	PW-3	PW-4	PW-5	PW-6	PW-7	PW-8	PW-9	PW-10
pH	7.7	7.5	7.62	7.77	7.65	7.54	7.68	7.99	7.89	7.56
EC (mS)	3.73	4.45	2.5	2.5	2.02	6.5	1.43	2.04	4.05	3.2
TDS (ppt)	1.55	2.3	1.5	1.32	1.24	3.89	1.4	1.25	2.3	2.2
DO (ppm)	4.4	4.2	4.1	4.4	4.7	4.3	4	4.2	4.2	4.4
Salinity (ppt)	1.89	2.68	1.5	1.8	1.78	4.5	1.9	1.23	2.03	1.89
Turbidity (NTU)	4.5	4.2	1.7	2.99	5.3	3.5	3.2	2.4	1.5	4.5
Hardness (mg/L)	68	70	50	48	40	53	42	58	62	33
O&G (mg/L)	2000	3500	1333.3	1000	2000	1100	1778.9	2360	2778.6	1356.8

3.2.2 Treatment of Physical and Chemical Contaminants of PW

The characterisation of the PW resulted in values which are much higher than the discharge regulation values according to WHO Standard. So, these values much be controlled before disposal.

HFM filtration of PW is performed to reduce its contaminants. Three membranes were used during the process.

- MF (0.1–0.4) µm
- UF (0.001–0.02) µm.
- NF 0.0001 µm.

In Figs. 14, 15, 16, 17, 18, 19, 20 and 21, values obtained from the parameters determined by water analyzer were seen, it can be inferred that the contaminants of PW of the samples were drastically reduced after the membrane filtration processes. The parameters that were drastically reduced were pH, salinity, EC, and turbidity, whereas DO in Fig. 5 shows good result after microfiltration membrane. Thus, from the above filtration processes it was seen that all the parameters of PW were within permissible limits by membrane filtration processes for safe disposal. Sample 7 shows high salinity, conductivity and TDS due to the presence of ions in the sample.

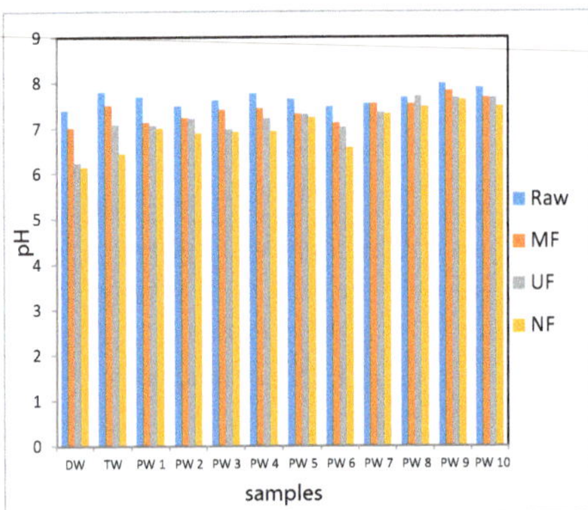

Fig. 14. Comparison diagram of pH after treatment

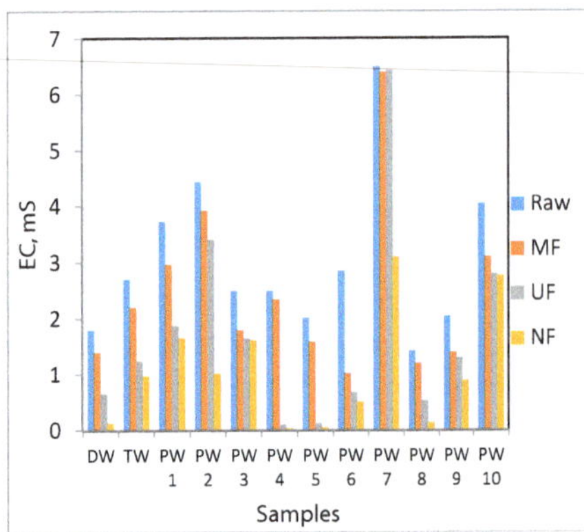

Fig. 15. Comparison diagram of conductivity after treatment

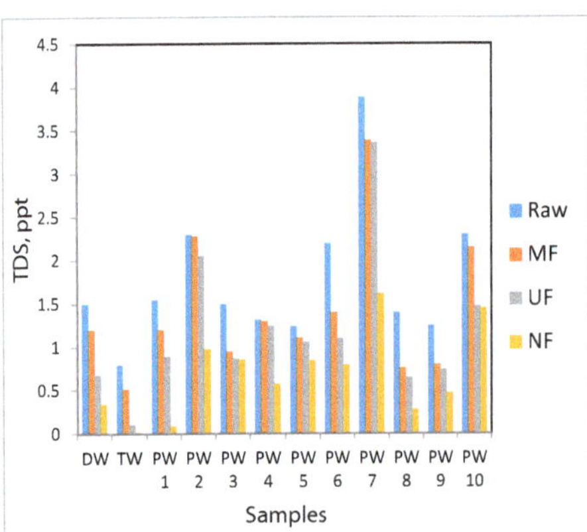

Fig. 16. Comparison diagram of TDS after treatment

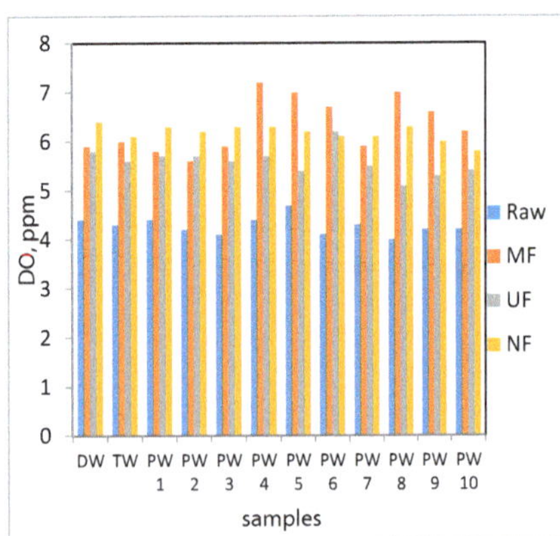

Fig. 17. Comparison diagram of DO after treatment

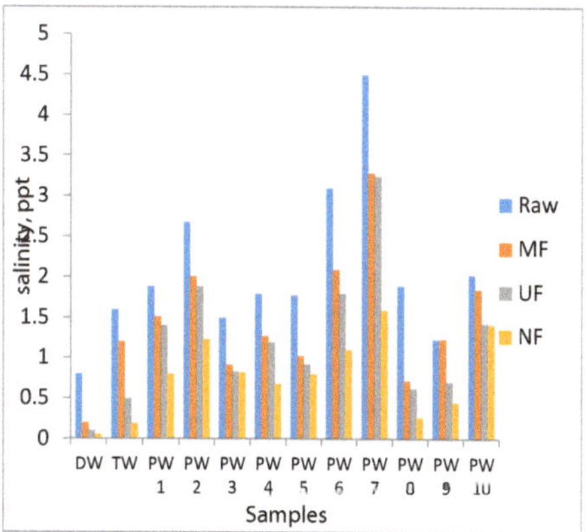

Fig. 18. Comparison diagram of salinity after treatment

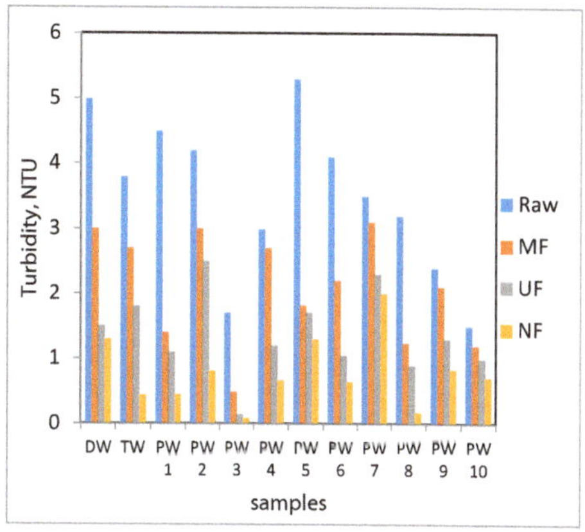

Fig. 19. Comparison diagram of turbidity after treatment

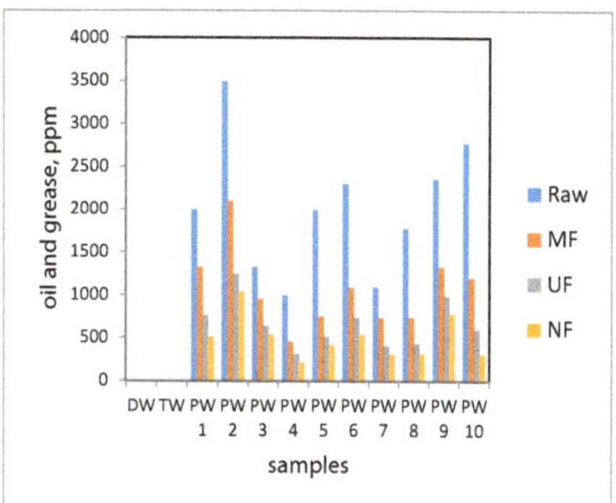

Fig. 20. Comparison diagram of oil & grease after treatment

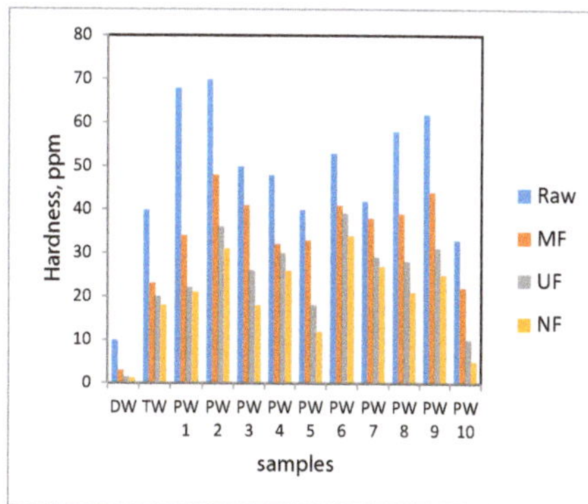

Fig. 21. Comparison diagram of hardness after treatment

3.2.3 Characterization and Treatment of Inorganics Present in PW

Table 8 shows the presence of inorganics in PW samples. The values are higher than the minimum discharge values of the WHO Standard and so it must be treated for its disposal.

The inorganic parameters obtained from the treated PW were analysed in a FP and AAS and were found to be drastically reduced in comparison to that of the raw samples of PW. From Figs. 22, 23, 24, 25, 26 and 27, it can be seen that NF process gives better result than other membrane treatment processes due to its smaller pore sizes.

Table 8. Inorganics determined by AAS and FP

Sl.No.	Parameters (ppm)	PW-1	PW-2	PW-3	PW-4	PW-5	PW-6	PW-7	PW-8	PW-9	PW-10
1.	Fe	2.47	1.011	0.024	0.337	0	0.312	1.236	2.13	0.216	0.056
2.	Cu	0	0	0	0	0	0	0	0	0	0
3.	Cr	0	0	0	0	0	0	0	0	0	0
4.	Mn	0.057	0.057	0.046	0.031	0.075	0.072	0.052	0.048	0.047	0.039
5.	Na	437.8	613.2	400.1	323	396.6	422.3	515.3	368.9	392.4	435.6
6.	K	26.3	44.2	19.46	8.74	14	32.5	10.2	7.1	22	9.2
7.	Ca	982	379	56.6	446.9	513.4	522	98	21	141.9	135.6
8.	Li	4.4	5.3	3.2	2.99	2.4	2.2	4.8	4.6	4.7	4.2

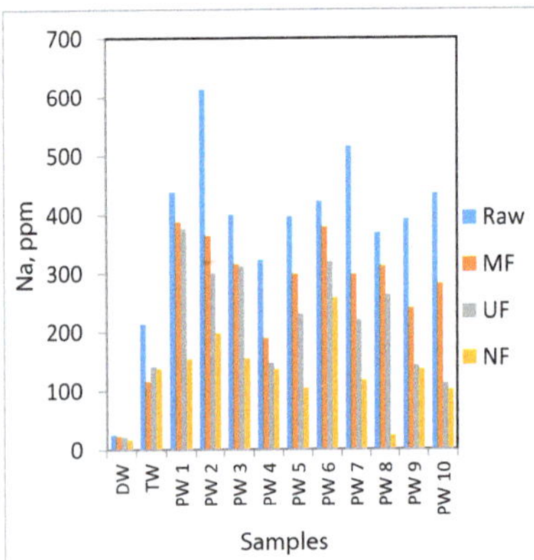

Fig. 22. Comparison diagram of Na after treatment

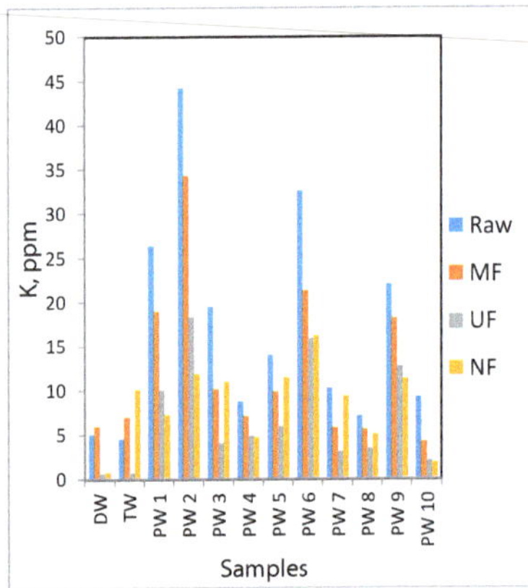

Fig. 23. Comparison diagram of K after treatment

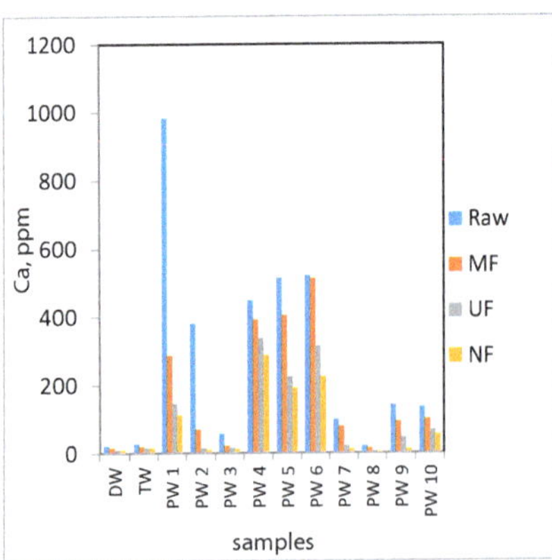

Fig. 24. Comparison diagram of Ca after treatment

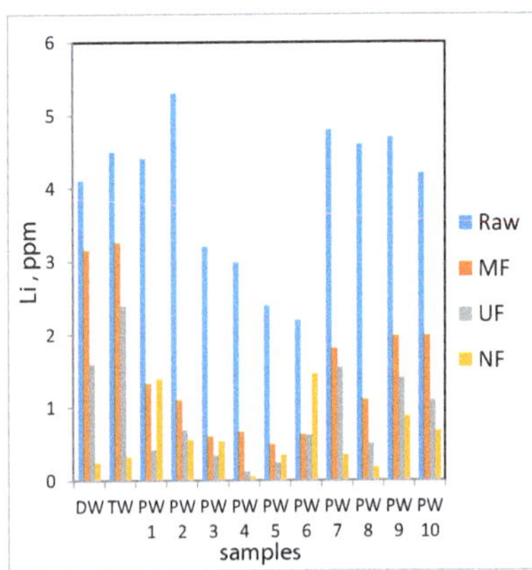

Fig. 25. Comparison diagram of Li after treatment

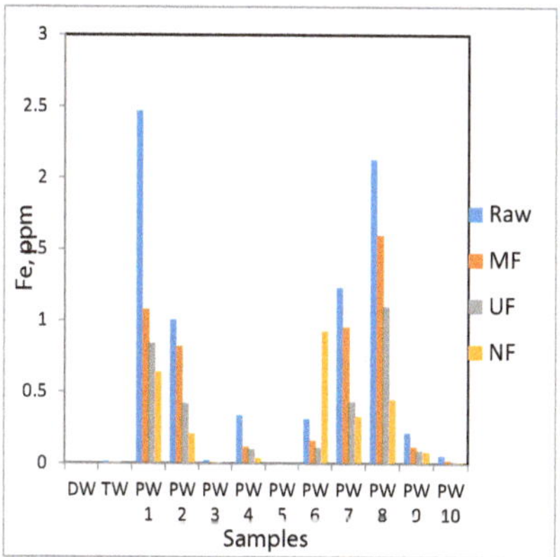

Fig. 26. Comparison diagram of Fe after treatment

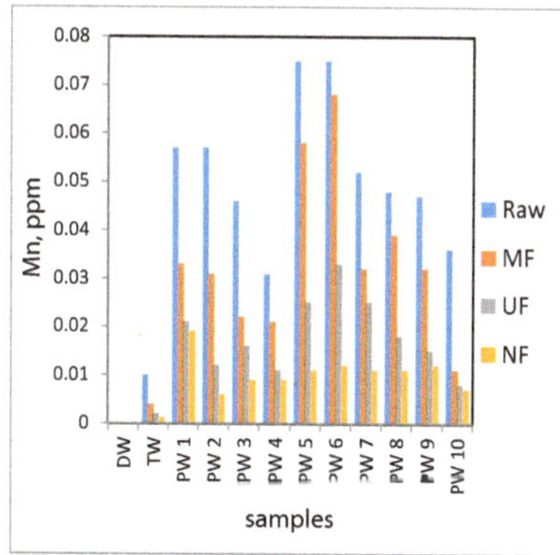

Fig. 27. Comparison diagram of Mn after treatment

3.2.4 Membrane Fouling

3.2.4.1 Microfiltration

Usually with an increase of pressure the flux must increase. From Fig. 28, it can be inferred that tape water and distilled water is showing the regular trend of increase in flux with increase of Trans-Membrane Pressure (TMP). In PW curves it can be seen that even after a certain TMP, the rate of increase of flux decreases very slowly, it means membrane fouling was not developed. The TMP must be increased in order to obtain the same amount of flux. This point of decrease of the rate of flux is known as the point of membrane fouling. Membrane fouling results in a decrease in performance of a membrane, caused by the deposition of suspended or dissolved solids on the external membrane surface, on the membrane pores, or within the membrane pores.

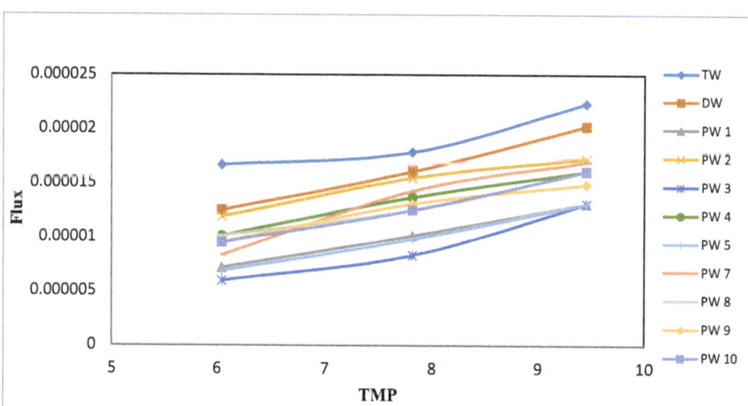

Fig. 28. Membrane fouling effect in micro-filtration process

Figure 29, shows that the volume collected after passing through the membrane did not decrease with the increasing of time that means membrane fouling was not developed during the treatment process.

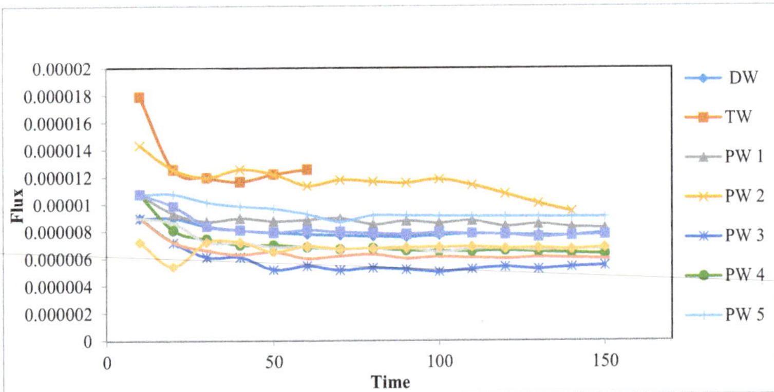

Fig. 29. Flux vs Time graph in microfiltration process

3.2.4.2 Ultrafiltration

In case of UF process, the volume that was collected was less than that of microfiltration process due to which flux get reduced for the same TMP and thus fouling effect occurs. From Figs. 30 and 31 it is clearly observed that there is no occurrence of membrane fouling in UF process as flux is increasing with the increase of TMP. The increase of flux with TMP (Fig. 30) clearly indicates that the membrane is not fouled and its pore spaces are not blocked.

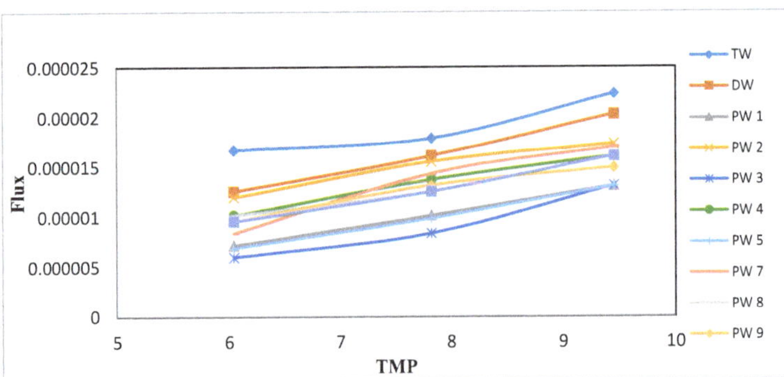

Fig. 30. Membrane fouling effect in ultra-filtration process

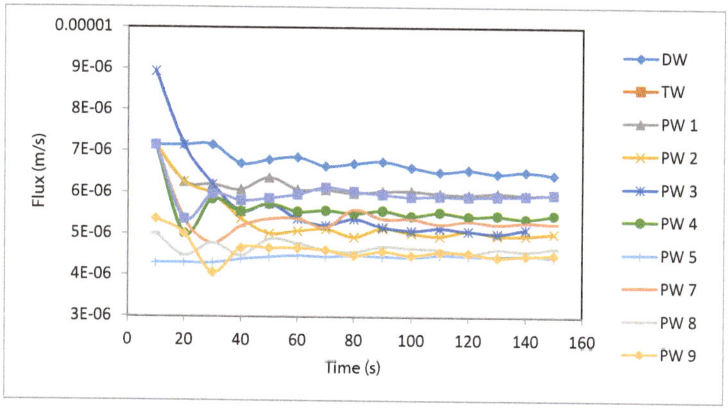

Fig. 31. Flux vs Time graph in ultrafiltration process

3.2.4.3 Nanofiltration

The volume collected in NF membrane is less in comparison to MF or UF membrane. The pore sizes are small in NF membrane and it showed high inclination to membrane fouling. In our study, we can infer from the Figs. 32 and 33 that there is no substantial fouling effect as flux is increasing with TMP and the flow rate is also increasing with time.

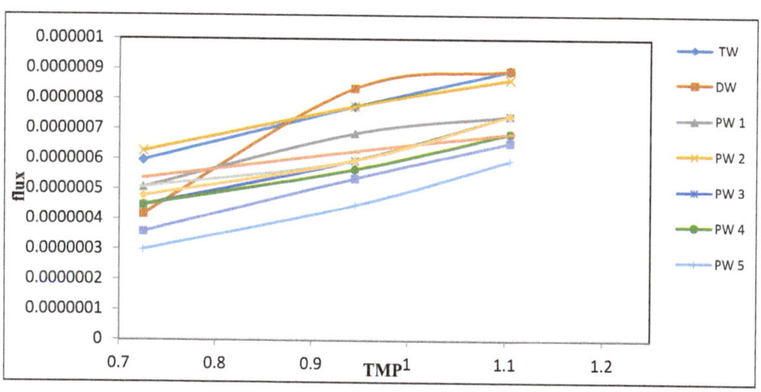

Fig. 32. Membrane fouling effect in nano-filtration process

3.3 Validation of Results

The results were validated in the Lenntech software. The Ryznar stability index (Table 3) and Langelier saturation index (Table 4) were calculated. Both the indexes shows that the PW samples are highly corrosive and it has a tendency to form scale. According to Langelier saturation index, PW-1 and PW-2 is supersaturated with respect to calcium carbonate ($CaCO_3$) and scale forming may occur whereas PW-3 to PW-10 is undersaturated with respect to calcium carbonate. Undersaturated water has a tendency

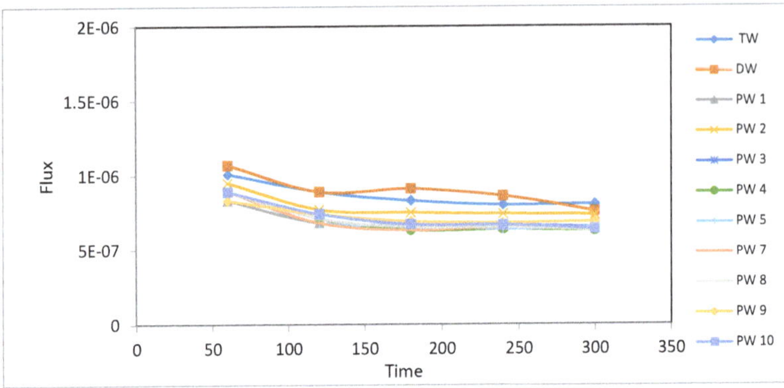

Fig. 33. Flux vs Time graph in nano-filtration process

to remove existing calcium carbonate protective coatings in pipelines and equipment. According to Ryznar stability index all the PW samples are highly aggressive. These PW samples actively gain minerals and metals, resulting in the corrosion of iron and copper surfaces and dissolving of concrete and stone surfaces.

4 Conclusion

The PWs collected from the oil fields of Upper Assam Basin were treated by using adsorption technology and HFM treatment processes. This report examines the physico-chemical and inorganic characterizations of the PW and the comparison are done between untreated PW and treated PW and their values were compared with the WHO specification for safe disposal of water into the environment. It is inferred that some of its parameters have higher values than the safe limits which can be adversely affect the environment. EC, turbidity, hardness, Na and Ca showed values much higher than the safer limits for its disposal whereas pH, salinity, DO and TDS gives satisfactory results. This experiment proves that sample 1 & 8 shows high salinity, EC and TDS due to the presence of Na ion. PAC adsorption is an effective means for reducing organic chemicals, chlorine, lead, and unpleasant tastes and odors from PW. After treating PW samples using PAC adsorption technology, there is a huge reduction in the values obtained from the parameters of untreated PW to that of the treated PW. The collected samples are also treated with HFM using micro-membrane, ultra-membrane and nano-membrane. All the membranes showed optimum efficiency regarding their filtration processes with respect to their pore sizes. Some of the parameters showed drastic decrease in the values whereas some showed slight decrease. UF gives better result than the other two membranes. The samples were diluted to increase the quantity of PW samples. Due to this dilution there are less contaminants in PW, resulting in the formation of a thin film of suspended or dissolved solids on the membrane surface. This thin film does not affect the performance of the membrane which results in negligible fouling of the membrane. Therefore, the water obtained by these treatment techniques of PW collected from the Upper Assam Basin, India can be re-used in agriculture, re-injection

Acknowledgement. We would like to acknowledge the following:

1. Indo-Tunisia Project no. DST/INT/TUNISIA/P-02/2017 entitled, "**Characterisation of oil field water to mitigate a major environmental hazard with recovery of usable water**".

2. Indo-US Project F. No. 194-1/2009(IC) dated February, 2015 entitled, "**Foam-assisted CO2 Flooding for the depleted Reservoirs of Upper Assam Basin and in Candidate Reservoirs in Louisiana**" of 21st century knowledge initiative program under University Grants Commission.

3. DBT Twinning Project No. BT/485/NE/TBP/2013 entitled "**Remediation of produced water from NE oil field by microbial and membrane intervention with modeling of an effluent unit for the recovery of usable water**".

4. Dibrugarh University for the laboratory facilities.

Nomenclature

AAS Atomic Absorption Spectrophotometer

BOD Biochemical Oxygen Demand

Ca Calcium

$CaCO_3$ Calcium Carbonate

Cr Chromium

Cu Copper

DO Dissolved Oxygen

EC Electrical Conductivity

EDTA Ethylene Di-Amine Tetraacetic acid

Fe Iron

HCO_3 Bicarbonate

HCl Hydrochloric Acid

HFM Hollow fibre membrane

KCl Potassium Chloride

K Potassium

KOH Potassium Hydroxide

Li Lithium

Li_2CO_3 Lithium Carbonate

MF Microfiltration

Mn Manganese

Na Sodium

NaCl Sodium Chloride

PAC Powdered Activated Charcoal

TDS Total Dissolved Solids

TMP Transmembrane Pressure

UF Ultrafiltration

WHO World Health Organization

References

1. Du, Y., Guan, L., Liang, H.: Advances of produced water management. In: Canadian International Petroleum Conference, 7–9 June, Calgary, Alberta (2005)
2. Guerra, K., Dahm, K., Dundorf, S.: Oil and gas produced water management and beneficial use in the Western United States. U.S. Department of the Interior Bureau of Reclamation, Science and Technology Program Report No. 157, Technical Service Center, September 2011
3. Khatib, Z., Verbeek, P.: Water to value – produced water management for sustainable field development of mature and green fields. J. Pet. Technol. **55**, 26–28 (2003)
4. Gogoi, S.B., Tamuli, A.S.: Recent advances in treating oil field produced water: a review. J. Pet. Eng. Technol. **4**(2), 366–370 (2014)
5. Veil, J.A., Pruder, M.G., Elcock, D., Redweik, R.J.: A White Paper Describing Produced Water from Production of Crude Oil, Natural Gas, and Coal Bed Methane prepared by Argonne National Laboratory for U.S. Department of Energy. National Energy Technology Laboratory (2004)
6. Al-Haleem, A.A., Abdulah, H.H., Saeed, E.A.: Components and treatments of oilfield produced water. Al-Khwarizmi Eng. J. **6**(1), 24–30 (2010)
7. Patel, H., Borgohain, D., Jena, S.: Hydro geochemistry of upper Assam oil field. J. Pet. Technol. **2**, 11–14 (2003)
8. Stone, B.: Management of Produced Water, pp. 37–41. Oxford Publications, Washington (2003)
9. Joel, O.F., Amajuoyi, C.A., Nwokoye, C.U.: Characterization of formation water constituents and the effect of fresh water dilution from land rig location of the Niger Delta, Nigeria. J. Appl. Sci. Environ. Manag. **14**(2), 37–41 (2010)
10. Jacobs, R.P.W.M., Grant, E., Kwant, J., Marqueine, J.M., Mentzer, E.: The composition of produced water from shell operated oil and gas production in the North Sea. In: Ray, J.P., Englehart, F.R. (eds.) Produced Water, Technological/Environmental Issues and Solutions, pp. 13–21. Plenum Press, New York (1992)
11. Konwar, D., Gogoi, S.B., Barman, J., Gogoi, M.P.: Correlation analysis of formation water of upper Assam Basin for a greener ecosystem. Int. J. Eng. Technol. Sci. Res. (IJETSR) **4**(8) (2017). ISSN 2394-3386
12. Gogoi, S.B., Sen, R.K., Rajbongshi, A., Hazarika, K.: Characterization of oil field produced waters of upper Assam Basin, India. Int. J. New Technol. Sci. Eng. **2**(1) (2015). ISSN 2349-0780
13. Kale, V.S.: Consequence of temperature, pH, turbidity and dissolved oxygen water quality parameters. Int. Adv. Res. J. Sci. Eng. Technol. **3**(8) (2016). ISO 3297:2007 Certified

14. Ramya, P., Jagadeesh Babu, A., Tirupathi Reddy, E., Venkateswara Rao, L.: A study on the estimation of hardness in ground water samples by EdtaTritrimetric method. Int. J. Recent Sci. Res. **6**(6), 4505–4507 (2015)
15. Barman, J., Gogoi, S.B., Sarmah, S.: Characterization of oil field formation water and its effect on tap water dilution for eco-friendly environment. Int. J. Eng. Technol. Sci. Res. **4**(8), 324–334 (2017). ISSN 2394-3386
16. Siong, Y.K., Idris, J., MazarAtabaki, M.: Performance of Activated Carbon in Water Filters. Water Resources java script (2013)
17. Rafferty, K.: Scaling in geothermal heat pump systems, U.S. Department of Energy, July 1999
18. Metcalf and Eddy, Wastewater Engineering Treatment and Reuse (2003)

The Effect of Helix Piles-Raft Foundation on Elastic and Consolidation Settlements in Tropical Fibrous Peat Soil

Ardy Arsyad[1(✉)] and Indra H. Ratu[2]

[1] Department of Civil Engineering, Faculty of Engineering, Hasanuddin University, Makassar, Indonesia
ardy.arsyad@unhas.ac.id
[2] Department of Civil Engineering, Faculty of Engineering, UKIP, Makassar, Indonesia

Abstract. This study investigates the performance of helix piles-raft foundation in tropical fibrous peat soil. Elastic bearing capacity of the foundation is represented by its stiffness to soil. Hence, we conducted the examination of helix pile-raft foundation stiffness to peat soil by using several analysis of Randolph and Wroth (1978), Richart et al. (1970), and Poulos-Davis (1980)–Randolph (1994). The analyses were adjusted and modified in order to accommodate the distinctive characteristics of helix piles such as radius and number of helices. The effectiveness of helix piles-raft foundation in reducing elastic and consolidation settlement in peat soil was examined. The results reveal that helix piles-raft foundation would reduce elastic and consolidation settlements. The number and size of helices affect significantly the stiffness of helix piles. Critical number of helix piles under raft foundation is found, depending on helices radius. Beyond this number, the helix piles would not affect the settlements of helix pile-raft foundation system significantly.

Keywords: Helix piles · Elastic and consolidation settlements
Stiffness · Fibrous peat soil

1 Introduction

Fibrous peat soils have become problematic ground for road infrastructure development in Indonesia since they have very low shear strength, high compressibility and high water content. Peat soil has primary compressibility at about 5 to 20 times to that in soft clay, while its secondary compressibility ranging from 0.05 to 0.07 of compressibility index (Mesri and Ajlouni 2007). The permeability of peat soil is about 1000 times permeability of soft clay (Mesri and Ajlouni 2007). The study of Kazemian et al. (2011) found that peat soil has undrained shear strength at around 3 kPa to 15 kPa. Having these conditions, construction of road pavement in peat soil would be unstable with excessive differential settlement. Several techniques have been implemented in reinforcement and improvement peat soil for road pavement such soil replacement, piled geotextile, and pile-slab. However, those techniques would be not cost effective particularly for deep peat soil where are dominantly found in Sumatera and Borneo.

H. Shehata and H. Poulos (Eds.): GeoMEast 2018, SUCI, pp. 246–256, 2019.
https://doi.org/10.1007/978-3-030-01917-4_20

Therefore, the study investigates the performance of alternative technique of combining helix piles with raft foundation in peat soil.

2 Method

The system of helix piles-raft foundation is similar to the system of piled raft foundation. Helix piles are connected to raft. Load on raft will be transferred partly to the helix piles and partly to ground. Helix piles will work to reduce settlement and differential settlements of raft, while raft will work to support structure load.

2.1 Determining of Stiffness of Helix Pile to Peat Soil

Stiffness of helix pile can be estimated through the analysis of pile head settlement under axial loading similar to the method of Randolph and Wroth (1978), It is assumed that the pile is a single pile in a homogeneous soil with a shear modulus increasing linearly with depth. However, in the case of helix pile, the analysis of Randolph and Wroth (1978) can be modified to deal with the characteristics of helix pile. Due to axial loading, soil below the helical plate (wb), soil along the cylindrical shear area (ws-1) and shaft pile will deform (ws-2) (Fig. 1b). For shallow helix pile, shaft resistance (PS-2) is too small compared to others resistance (Mohajerani et al. 2016), so that deformation along the shaft pile (ws-2) does not significantly contribute to pile settlement. Therefore, pile head settlement for helix pile can be determined based on the deformation of soil below helical plate (wb) and the deformation of soil along cylindrical shear failure (ws-1).

Deformation of the soil surrounding the helix pile is typically similar to shearing of concentric cylinders introduced by Randolph and Wroth (1978). The vertical equilibrium of an element of soil is given by Eq. 1.

$$\frac{\delta}{\delta x}(r\tau) + r\frac{\delta \sigma_z}{\delta z} = 0 \tag{1}$$

Where:

r = horizontal distance to the pile axis
τ = shear stress
x = horizontal coordinate
σ_z = vertical stress
Z = depth

Shear stress generated by pile loading is much higher than vertical stress, therefore shear stress in the soil surrounding the helix pile is following:

$$\tau = \frac{\tau_0 r_0}{r} \tag{2}$$

r_0 = radius of the pile shaft.
τ_0 = shear stress at the pile shaft.

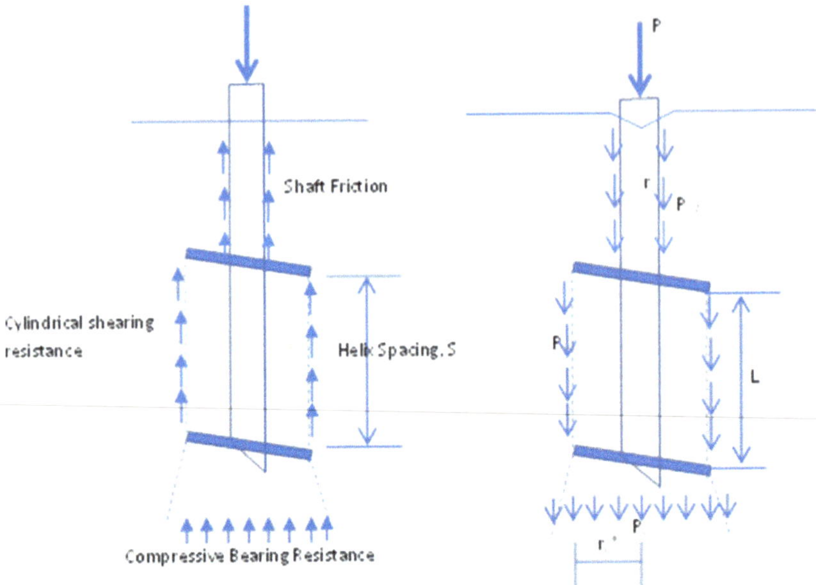

Fig. 1. Bearing capacity of helix pile under a compressive load, deformation patterns of helix pile in the upper and lower soil layers.

As the helix pile has cylindrical shear and pile shaft stress, therefore, Eq. 2 can be modified as:

$$\tau = \frac{\tau_0 r_0}{r} + \frac{\tau_0^* r_0^*}{r^*} \tag{3}$$

r_0^* = radius of cylindrical helix shaft
τ_0^*= shear stress at the cylindrical helix failure.

Shear strain of the soil can be determined with the Eq. 4.

$$\gamma = \frac{\tau}{G} = \frac{\delta u}{\delta z} + \frac{\delta w}{\delta r} \tag{4}$$

Displacement yielded by shear stress is:

$$w_s = \int_{r_0}^{\infty} \gamma\, dr + \int_{r_0^*}^{\infty} \gamma\, dr^* = \frac{\tau_0 r_0}{G} \int_{r_0}^{\infty} \frac{dr}{r} + \frac{\tau_0^* r_0^*}{G} \int_{r_0}^{\infty} \frac{dr}{r^*} \tag{5}$$

For shallow helix pile, pile shaft resistance is too small compared to cylindrical shearing resistance (Mohajerani et al. 2016) and its displacement can be ignored (Eq. 6).

$$w_s = \frac{\tau_0^* r_0^*}{G} \int_{r_0}^{\infty} \frac{dr}{r^*} \tag{6}$$

If radius influence (rm) of Randolph and Wroth (1978) is employed as upper boundary of the shearing influence (the Eq. 7).

$$w_s = \frac{\tau_0^* r_0^*}{G} ln\left(\frac{r_m}{r_0^*}\right) = \zeta \frac{\tau_0^* r_0^*}{G} \tag{7}$$

where:

$$\zeta = ln\left(\frac{r_m}{r_0^*}\right)$$

The factor ζ is a relation between the radius of influence of the helix pile and the radius of the helix plate.

For compressible pile, cylindrical shaft settlement varies with depth. Shear stress is no longer assumed as constant along the pile (Eq. 8).

$$w_s(z) = \zeta \frac{\tau_0^*(z) r_0^*}{G} \tag{8}$$

The helix pile is considered as elastic, and its compressive strain in depth may expressed with transmitted load.

$$\frac{\delta w(z)}{\delta z} = -\frac{P(z)}{\pi r_0^{*2} E_p} = -\frac{P(z)}{\pi r_0^{*2} \lambda G} \tag{9}$$

Soil – helix pile stiffness ratio $\lambda = E_p/G_L$

The transmitted load is related to the variable shear stress on the shaft surface.

$$\frac{\delta P(z)}{\delta z} = -2\pi r_0^* \tau_0^*(z) \tag{10}$$

$$\frac{\delta^2 w(z)}{\delta z^2} = \frac{2}{r_0^2 \zeta \lambda} w(z) \tag{11}$$

Solution:

$$\mu L_c = \frac{L_c}{r_0^*}\sqrt{\frac{2}{\zeta\lambda}}$$

$$\frac{P_t}{G r_0^* w_t} = \left[\frac{4}{\eta(1-v)} + \frac{2\pi L_c tanh(\mu L_c)}{\zeta r_0^* \mu L_c}\right]\left[1 + \frac{4 L_c tanh(\mu L_c)}{\eta(1-v)\pi\lambda r_0^* \mu L_c}\right]^{-1} \tag{12}$$

The fraction of load taken by base of helix pile is shown in Eq. 13.

$$\frac{P_b}{P_t} = \left[cosh(\mu L_c) + \frac{\pi(1-v)L_c\eta sinh(\mu L_c)}{2\zeta\mu L r_0^*}\right] \quad (13)$$

2.2 Determining the Stiffness of Raft Foundation

The stiffness of raft was computed based on Richart et al. (1970) equation

$$K_r = \frac{G_s}{(1-v_s)}\beta_z\sqrt{(4cd)} \quad (14)$$

where: G_s = the shear modulus of the soil; v_s is the Poisson's ratio of the soil; and t = the thickness of the raft, and c and d are coefficient raft dimension.

2.3 Determining Stiffness of Helix Piles-Raft Foundation

Load sharing between the raft and the helix piles was computed using simple method Randolph (1994), and Poulos and Davis (1980), shown in Eq. 15.

$$K_{pr} = \frac{1-0.6\left(\frac{K_r}{K_{hp}}\right)}{1-0.64\left(\frac{K_r}{K_{hp}}\right)}K_{hp} \quad (15)$$

where k_{pr} = stiffness of piled raft; K_{hp} = stiffness of the helix pile group; k_r = stiffness of the raft alone, and β_{hp} = raft – pile interaction factor.

The proportion of the total applied load carried by the helix pile is in Eq. 16.

$$\beta_{hp} = \frac{1}{1+a} \quad (16)$$

$$a = \frac{0.2}{1-0.8\left(\frac{K_r}{K_{hp}}\right)}\left(\frac{K_r}{K_{hp}}\right) \quad (17)$$

Tri-linear curve was generated by computing mobilization of the helix pile capacity according Eq. 18.

$$P_1 = \frac{P_{up}}{\beta_{hp}} \quad (18)$$

$$P_{hp} = \beta_{hp}P \le P_{up} \quad (19)$$

$$P_r = P - P_{hp} \quad (20)$$

where P_{hp} = load on helix piles, P_r = load on the raft, P_{up} is the ultimate bearing capacity of helix pile group.

3 Results

The effect of radius of helices on the stiffness of helix pile has been examined, and the result is shown Fig. 2. The increase of helices radius would increase the stiffness of the helix pile. The number of helices in a helix pile affects the stiffness of helix pile (Fig. 3). A commercial software of Deepexcavation (2017) was employed to examined the performance of the helix pile. It must be noted that the software computes the bearing capacity and settlement of helix pile based on Vesic formula (1974). It was found that the stiffness of the helix pile obtained from Modified Randolph and Wroth is a bit overestimate than that obtained from the software, about 5%–12%. This is due to such disturbance factor is ignored in the analysis. However, the result is well agreement particularly for helix pile with small radius helice, $r_0^* < 0.3$ m (Fig. 3).

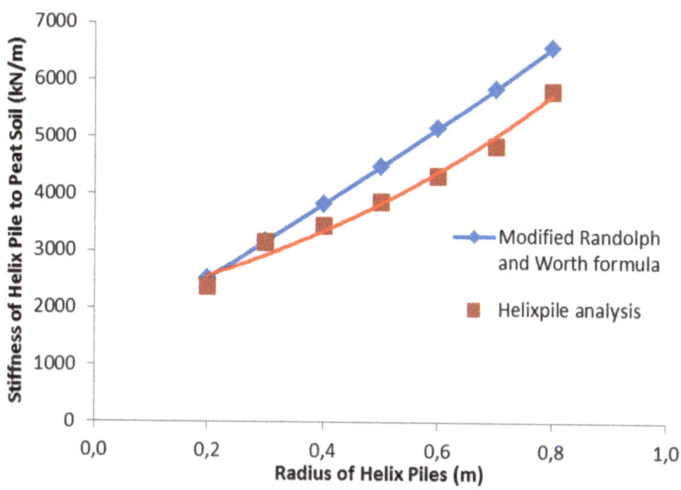

Fig. 2. Effect of radius of helice on helix pile stiffness.

3.1 The Effect of Helix Piles-Raft Foundation on Settlement

A segment of raft with dimension of 11 m × 10 m × 0.3 m was subjected with 65 tons (Fig. 4). The raft lies on a peat soil with E of 318 kPa, v of 0.35 and C_u of 3.975 kPa, C_c of 1.83 and e_0 of 1.50 (Table 1). Bearing capacity of the raft in peat soil is 192.1 kPa at ϕ of 26°. However, the elastic settlement of the raft foundation is 474 mm. It is larger than the allowable settlement (60 mm). By using one-dimensional Terzaghi, and Boussineq and Fadum coefficients for stress distribution along the depth of the soil, the consolidation settlement is about 1,312 mm.

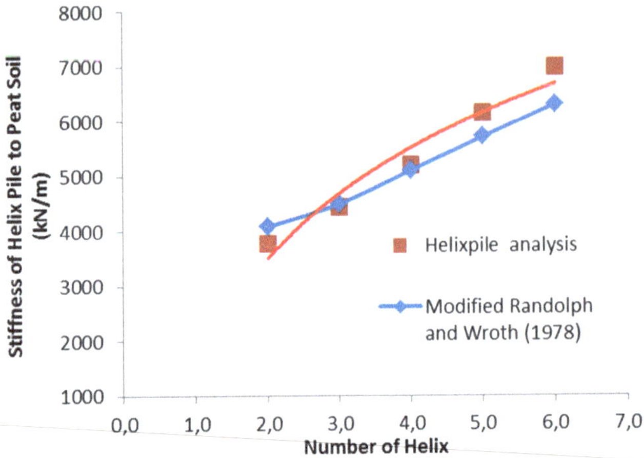

Fig. 3. Effect of number of helices on helix pile stiffness.

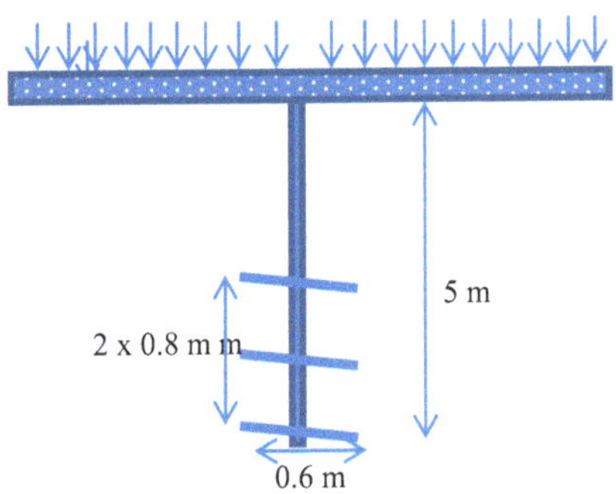

Fig. 4. Schematic helix piled raft on peat soil loaded with traffic load.

3.2 The Effect of Helix Piles Numbers on Settlement

The increase of number of helix piles under the raft foundation would decrease the elastic settlement of the foundation (Fig. 5). Critical number of helix piles is about 9 piles to 16 piles, particularly for helix pile with helice radius above 0.5 m. Beyond this numbers, the helix pile number would not affect the settlement. If the allowable elastic settlement required for road pavement is 60 mm, then the minimum required helix piles is in 16 piles with minimum helice radius at 0.5 m. The consolidation settlement of the helix pile - raft foundation is below 20 mm (Fig. 6). The fraction of load taken by helix piles would increase following the increasing number of helix piles (Fig. 7). It was found fraction of load taken by helix piles will increase from 47% to 97%.

Since the raft foundation settles excessively, more helix piles are needed to reduce the settlement. It can be seen in Fig. 8, the increase of helix pile numbers would decrease the percentage of load distributed to helix piles. So that, the load distributed to helix piles would not overpasses the helix piles capacity. For small helix piles (radius < 0.3 m), the

Table 1. Properties of peat soil

Weight volume of peat	$\gamma = 11.4$ kN/m^3	Al-Ani et al. (2013)
Undrained Shear Shear Strength in compression	0.5 to 0.75 to preconsolidation pressure. Cu = 3.975 kPa–5.963 kPa	Mesri and Ajlouni (2007)
Young Modulus	20–80 to undrained shear strength. E = 79.5 kPa–477 kPa	Mesri and Ajlouni (2007)
Compressbility index (Cc)	Cc = 1.83–2.74	Al-Ani et al. (2013)
Internal shear angle	$\phi = 40°–60°$ $\phi = 53°–57°$ $\phi = 3°–26°$	Mesri and Ajlouni (2007) Kazemian et al. (2011) Kuat et al. (2014)
Coefficient of consolidation (Cv)	Cv = 20–100 m2/year	Mesri and Ajlouni (2007)
Secondary compressibility index	Cα = 0.05 – 0.07 Cα = 0.06 Cc Cα = 0.03 – 0.04	Mesri and Ajlouni (2007) Al-Ani et al. (2013)

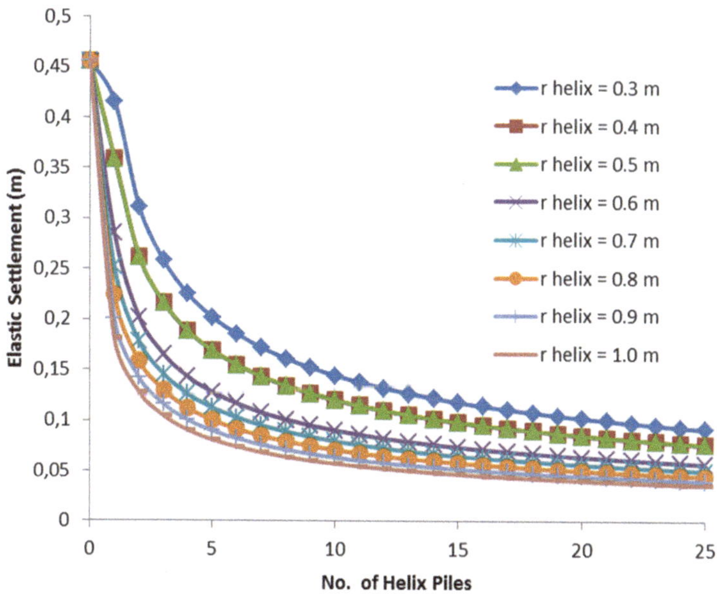

Fig. 5. Effect of number of helix piles on elastic settlement.

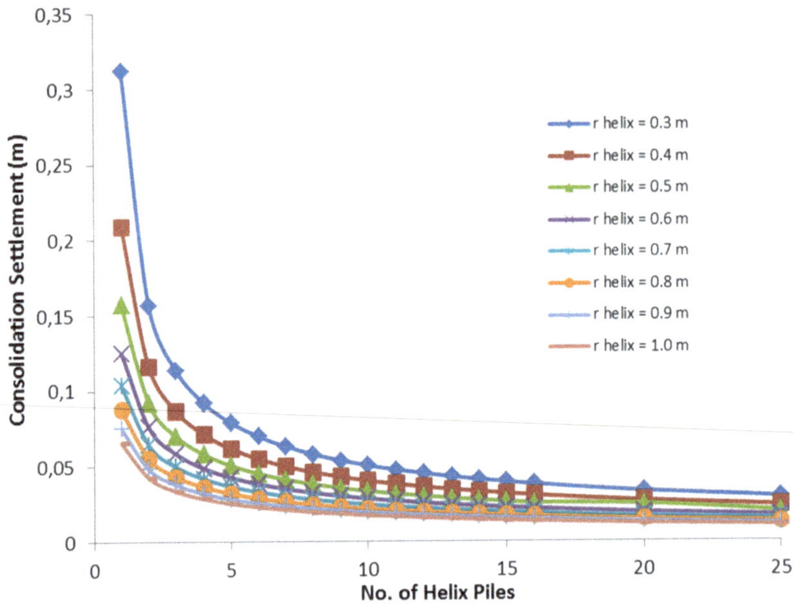

Fig. 6. Effect of number of helix piles on consolidation settlement.

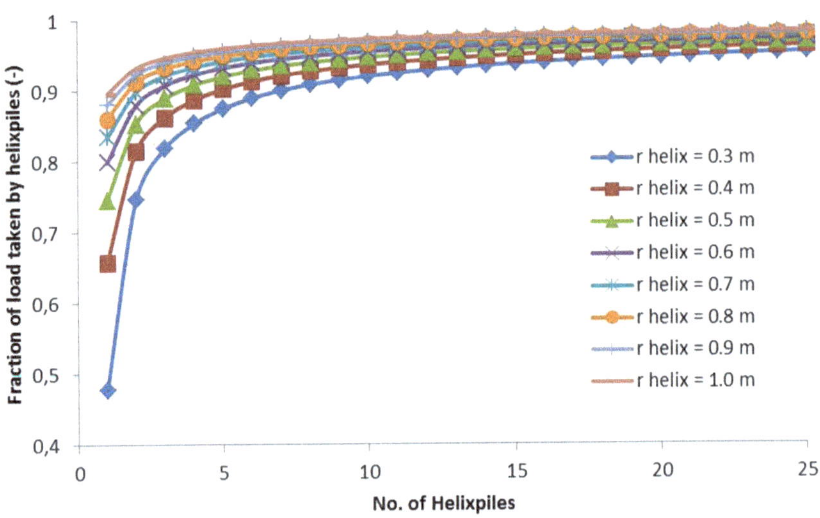

Fig. 7. Effect of number of helix piles on fraction of raft load taken by helixpiles.

helix piles could be be mobilized due to the distributed load has overpassed the helix pile capacities. Even, the helix pile numbers are maximum at 25 piles. On the other hand, for large helix piles (radius $\geq$ 0.4 m), the helix pile would not be mobilized, as long as the minimum 18 helix piles are installed under the raft foundation.

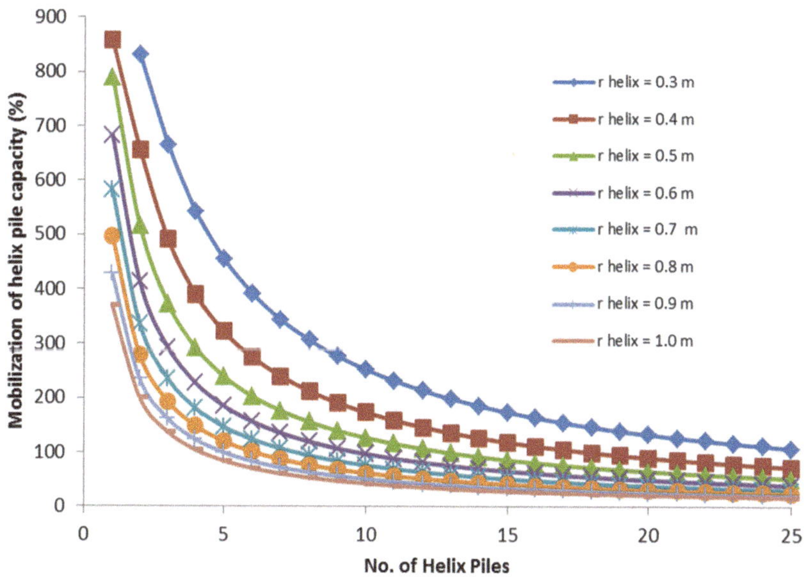

Fig. 8. Effect of number of helix piles on mobilization.

4 Conclusions

1. The increase of number of helix piles under raft foundation would decrease the elastic and consolidation settlement of the foundation system in peat soil subjected with traffic load.
2. The dimension and number of helice in helix pile would affect the stiffness of helix pile.
3. Critical number of helix piles under raft foundation is found at 10 piles to 15 piles, depending on helices radius. Beyond this number, the helix piles would not affect the settlements significantly.
4. The helix pile would not be mobilized, particularly for the helix pile with helice radius $\geq$ 0.4 m.

References

Al-Ani, H., Oh, E., Gary, C.: Characteristics of embedded peat in coastal environments. Int. J. GEOMATE **5**, 609–618 (2013)

Kazemian, S., Huat, B., Prasad, A., Barghchi, M.: A state of art review of peat: geotechnical engineering perspective. Int. J. Phys. Sci. **6**(8), 1974–1981 (2011)

Kuat, B.B.K., Prasad, A., Asadi, A., Kazemian, S.: Geotechnics of Organic Soils and Peat. CRC Press, Boca Raton (2014)

Mesri, G., Ajlouni, M.: Engineering properties of fibrous peats. J. Geotech. Geoenviron. Eng. **133** (7), 851–866 (2007)

Mohajerani, A., Bosnjak, D., Bromwich, D.: Analysis and design methods of screw piles: a review. Soils Found. **56**(1), 115–128 (2016)

Poulos, H.G., Davis, E.H.: Pile Foundation, Analysis and Design. University of Sydney, Sydney (1980)

Richart, F.E., Hall Jr., J.R., Woods, R.D., Richart, F.E.: Vibrations of Soils and Foundations. Prentice Hall, Englewood Cliffs (1970)

Randolph, M.F.: Design methods for pile groups and piled rafts. In: Proceedings 13th International Conference on Soil Mechanics and Foundation Engineering, New Delhi (5), pp. 61–82 (1994)

Randolph, M.F., Wroth, C.P.: Analysis of deformation of vertically loaded piles. J. Geotech. Geoenviron. Eng. 104(ASCE 14262) (1978)

Evaluation of the Technical Alternatives for the Stabilization of the Sublevel Layer in the BRT Sul Corridor Works, Brasília, Federal District, Brazil

Haroldo Paranhos[1(✉)], Rideci Farias[1], Marcos Costa[2], Joyce Maria Lucas Silva[3], and Roberto Pimentel[4]

[1] Reforsolo Engineering, Brasília, Brazil
reforsolo@gmail.com, rideci.reforsolo@gmail.com
[2] OAS, Brasília, Brazil
marcos.costa@oas.com
[3] CONCREMAT, Brasília, Brazil
joyce.civil@gmail.com
[4] University of Brasília, UnB, Brasília, Brazil
eng.robertopimentel@gmail.com

Abstract. As a public transport solution, the BRT concept is increasingly common in large cities. The city of Gama/DF, place of implantation of this modal is the result of a Geology characterized by a layer of Clay red latosoils, denominated of porous and collapsible clay. Bibliographical studies indicate that many paving works present exaggerated deformations when the lack of detailed characterization of these profiles, in front of the requests of the traffic. The use of dimensions based on the CBR theory and the verification of the quality of earthworks and pavement layers, do not present satisfactorily the effects of deformability of these structures, since the services of control of compaction affirm a degree of compaction compatible with the services, but when measuring the deformation by the Benkelman beam method, values are set below, to the extent that they are not accepted by the inspection. Important factors such as granulometry, Atebergs limits and the presence of hygroscopic clay minerals accentuate the problem, since these soils present hydration of water in contact with the medium and high capillarity, to the point that it is difficult to maintain optimal moisture during the services and the recompaction becomes impracticable, since the disorganization of macropores makes these soils even smaller from the texture point of view. Due to the degree of difficulty presented in working with these types of soils, and the tight deadline for the implementation of the compaction service stage, five solutions of subgrade reinforcement were proposed and tested in order to direct the most favorable technical and economic solution to the problem to the project schedule. These solutions transit through the uses of structural layers, with Rachão stone; subleito reinforced with microgrel; subleito reinforced with geotextile; compacted subgrade with compacted subgrade with 5% of virgin lime. All the solutions were evaluated in experimental stretches of approximately 100 m of extension, of the work itself, by means of the conventional technological control and with the measurement of the deformation with the use of the benkelman beam.

H. Shehata and H. Poulos (Eds.): GeoMEast 2018, SUCI, pp. 257–268, 2019.
https://doi.org/10.1007/978-3-030-01917-4_21

1 Introduction

This document presents the technical alternatives for the stabilization of subleito along the works of the BRTSul corridor, in the section Gama - Catetinho in due to non-compliance with the parameters acceptability of deflections by the Beam Benkelman. It should be noted that the earthwork in sub-stretch 1, stakes of 0 (zero) to 20, Range

Balloon LD to CAUB according to the basic assumptions established in executive procedures and also in compliance with the norms and project specifications and other requirements to the undertaking in question. In sequence, with the subsequent tests to determination of deflections in the regularization and reinforcement of the subgrade by the Benkelman beam, it was verified the non-compliance with the parameters results. Thus, in this document, the analysis of possible alternatives with a view to solution of the problem in question and thus technical considerations meet the necessary support capacity for the perfect performance of the floors designed.

2 Scope of Services

Firstly, the elements of projects (survey reports, type sections, results of laboratory ests, etc.);

In conjunction with the field visits, new polling campaigns are planned collection of eformed samples for characterization and determination of the moisture profile;

Then the alternatives were suggested techniques for stabilizing the subleito. At the same time new tests were for the definition of soil mixtures and lime.

Subsequently, five experimental sections, using the following solutions: layer of Ration, layer of gravel, soil reinforced with geotextile, soil lime and soil reinforced with microgrel.

These layers were evaluated for feasibility and deformability.

3 Information and Considerations on the Soil of Region

Such Geology/ Geomorphology is characterized by a layer of Oxisols and soils red clayey lateitic species, called porous and collapsible clay. This material is overlapping, in general, to soils of two distinct origins: a silt from shale change or a sequence of metasiltites and quartzites that geologists call them metarrhythmias. They belong to the Paranoá Pre-Upper Cambrian.

The climate alternates between a very dry and another rainy with 6 months of duration, leading to a lateralization process that consists of the leaching of soluble upper layers and deposition below. This one process leaves in the upper layers of clay a large amount of voids, or pores, resulting in high void indices, low specific gravity and high permeability.

In this way, extending the issue to the pavement area, it has been noticed that many of the floors in the Federal District have appreciable deformations that may also be related to regularization and reinforcement of the subgrade layer.

Another important issue to consider is that soils may show variations of the modulus of resilience depending on humidity, energy and the compaction method. Beyond In addition, after compaction, soil floor structure may vary with the climatic and hydrological conditions. THE resilient deformation increases significantly in the compacted soils on the humid branch - as in the In the case in question it was found that the field is above the optimum moisture compaction of soil - in relation to those compacted in the dry branch, with the same specific dry mass (Preussler 1983). Working with the high hypothesis resilience modules obtained by compaction soil on the dry branch may not be advisable, since it is difficult to ensure that there will be an increase in soil moisture life of the floor. Increased humidity may cause a significant drop in the module of resilience.

Compaction at low advantageous if there is a concomitant increase in compaction energy. The solos show a significant increase in the modulus of resilience with increasing compaction energy (Bernucci 1995). THE loss of soil moisture requires an increase in the resilience; can not be neglected in the However, soils losing moisture may contraction cracking. O produces a reduction of the equivalent resilience of the layer. It is important to note that sometimes in the paving community, there is a tendency to use simpler assays to estimate the value of the resilience module of materials. The authors advise against use of ISC correlations to estimate the value of resilience module since it does not there is a consistent relationship between these parameters - the first considers the material due to excessive permanent deformation, while the latter considers casualties elastic deformations.

Santos (1998) shows that the movement of the sesquioxides can be explained in terms of the soil water regime tropical. During the rainy season, saturated horizons and water movement in the soil is down, leaching the layers higher. In the dry season there is already intensive drying of the surface horizons and water movement is up.

The floors are not subject only to the traffic loads actions. The physical environment also influences his life. useful, and this should not be overlooked. These actions (climate and traffic) together constitute the factors that impose a limited to pavements (Medina 1997).

Motta (1991) states that climate is one of the factors that most affect the performance of a floors. Represented by its elements air temperature and precipitation, the climate influences the floors through various mechanisms, such as materials, the alteration of the resilient modules or the change of the equilibrium moisture. Villibor (1981), using a beam Benkelman, measured the deflections of the stretches analyzed, obtaining low values (between 10 and 70 hundredths of a millimeter), and concluded that if the moisture conditions of the base and sub-base, this behavior of low deflections should persists over many years in pavements with lateritic basis.

Elliot and Thompson (1985) and Elliot and Thornton (1988) report that most of the soils exhibit a decrease in the modulus of resilience when humidity increases, resulting in an increase in the deflections of compromise the service life of the floors.

According to Medina (1997), the nature of fraction determines the behavior of the resilience. When at the percentages by weight between the silt fraction and fraction (less than 0.074 mm) is less than 50% soil is less resilient.

Soils change their elastic behavior moisture content after compaction. There are more sensitive soils, the silts, which may vary, for the same condition of dry density, elastic behavior low to very high behavior, with a humidity range of 2 to 3 points. This means that it may be deformed elastic for a given load, but if its humidity rises 2% or 3%, absolute, it has very high elastic deformations. There is nothing that can be done to avoid this humidity fluctuation.

On the other hand, sands and boulders, without cohesion, maintain the elastic behavior low deformation, i.e. low deformation for any moisture variation.

For safety reasons, it is convenient to have the deflection values for the most adverse effects. How the deflections vary with the type of soil and the variation of moisture, it is difficult to obtain adjustment factors reliable. The DNIT (DNER) usually works with adjustment coefficients for two types of soil, sandy and loamy, and does not silt, which is more critical.

3.1 Local Pedology

According to Paranhos (1998), soils of the region are predominantly composed of Red Latosols and Red Yellow (Fig. 1). Therefore there are portions of Gleissolos ("hydromorphic soils") adjacent to highway. In general, the profile features a pack of sandy clay, red, very soft to soft, and SPT 2 to 4 blows.

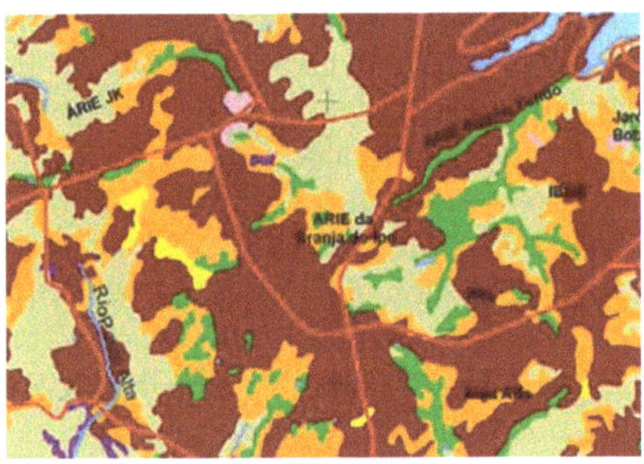

Fig. 1. Pedological map of the study area (Embrapa Cerados 2004).

3.2 Applied Pedology

The interpretations of the surveys contained in the Geological/Geotechnical report of the work in question confirm the existence of the characteristics of the main groups of soils found: Latosols and hydromorphs.

For the latossolos the probing detected thick and mature layers, of clay character, of low resistance to penetration, together with a high porosity, in the more superficial horizons, with this, the geotechnical problem attributed to its structural characteristic, the phenomenon of collapsivity, that is to say, its abrupt density by the collapse of the structure, when saturated, even without being subjected to an additional loading. The saturation in water not only reduces the resistance of the latosols to the erosion, but, mainly, causes repression in shallow foundations by the structural collapse of the soil.

With regard to Water Level, in general, according to the surveys carried out, the water table where it was found is very shallow, between the Gama and Park Way, varying from 0.82 m to 16.00 m, this being for the latosols and probably for the hydromorphs.

4 Location of the Collection of Samples and Results of the Essay

Figure 2 shows the locations where collected the 5 (five) soil samples along the stretch Range - Catetinho, to the performance of the compaction tests and the natural moisture.

Fig. 2. Micro view of the sample collection area.

4.1 Compaction Test

The purpose of the compaction tests is to determine the optimum moisture and the dry density maximum. Table 1 shows the results of the compaction tests.

Table 1. Optimum moisture content and maximum dry mass of collected soils.

Amostra	1	2	3	4	5
Localização (Estaca)	Est. 10 + 00	Est. 25 + 00	Est. 39 + 00	Est. 92 + 00	Est. 42 + 00) CAUB-Catetinho
Umidade Ótima (%)	28,2	28,9	23,5	27,49	28,2
Massa Específica Seca máxima (kN/m^3)	17,60	17,4	18,05	17,2	15,66

4.2 Humidity Profile

As a way to verify the natural humidity of the soil, data were collected on the variable depths from 0.50 m to 1.50 m. The Fig. 3 shows the moisture profiles for the studied. It can be seen that 75% of natural soil moisture levels are above the optimal compaction humidity for the depths studied.

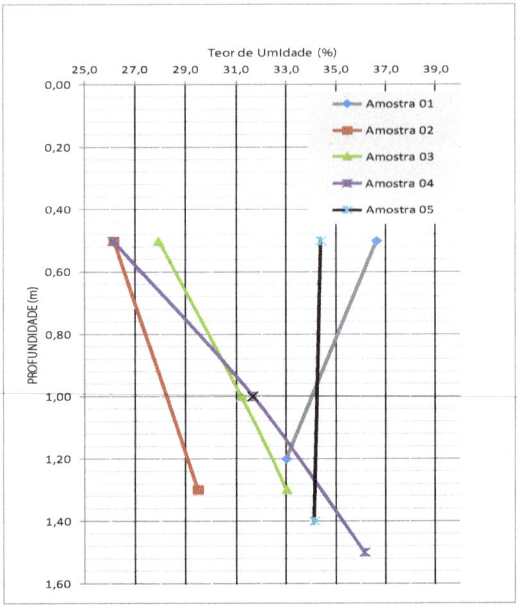

Fig. 3. Moisture profiles at the collection points of the samples.

5 Proposition of Possible Solutions to the Problem in Question

Among the various alternatives for strengthening layer, such as that of the subgrade, there is replacement of the existing soil by another better quality, reinforcement by "needling", with piles, reinforcement with synthetic material (geosynthetics), soil improvement with additives, such as lime and (or) cement, between solutions on the market. Next feasible alternatives to the project in question.

5.1 Reinforcement with Stony Material

In the process sequence, the granular material was spread and compacted, with the surface grit 0 (Figs. 4 and 5).

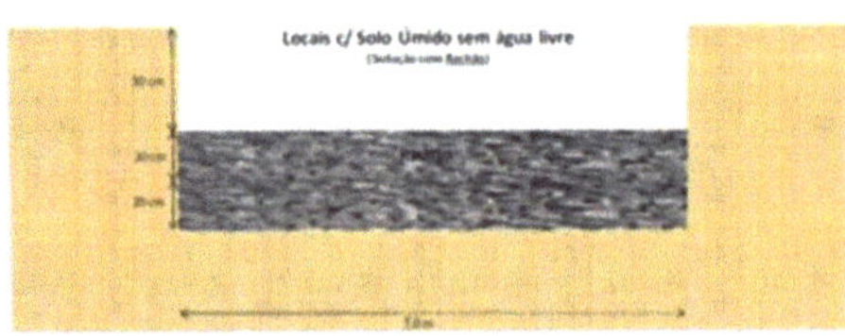

Fig. 4. Crack layer type section.

Fig. 5. Crack layer type section.

5.2 Replacement of Existing Soil (Gravel)

Figures 6 and 7 present the solution proposal concerning the replacement of land by gravel.

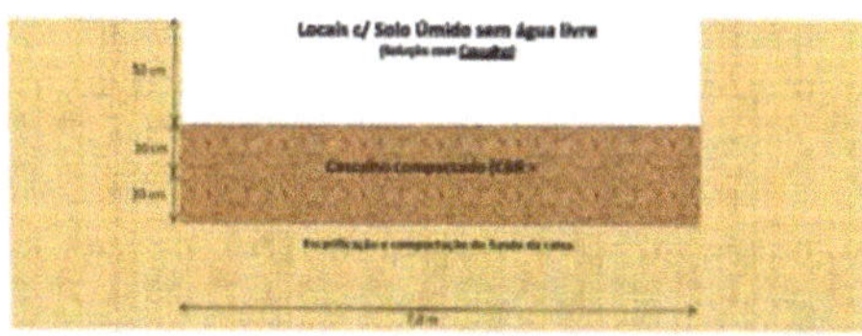

Fig. 6. Solution type with gravel.

5.3 Reinforcement with Geotextile and Sand

Figures 8 and 9 show details of the solution with geotextile and sand.

The purpose of using the geotextile is incorporate a layer of reinforcement in the soil, to promote the separation between the bottom soil of the cava and the layer of sand. The purpose of sand is to promote the capillarity the underlying layers.

5.4 Reinforcement with Microgress

The purpose of using the geogrid is incorporate a reinforcement layer to the soil. Figures 10 and 11 show the solution proposal concerning the introduction of micro-as a structuring element of the layer compacted.

5.5 Addition of Lime to Soil

The objective of adding Cal to the soil is to improvement of the layer's carrying capacity and the possibility of reducing soil moisture for optimal moisture through addition of virgin lime.

Fig. 7. Application of lateritic gravel.

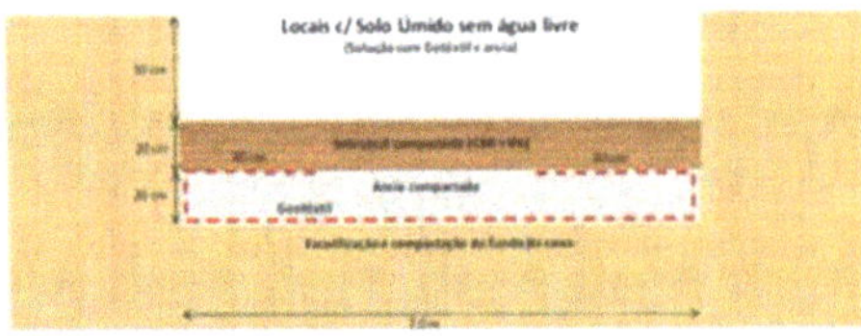

Fig. 8. Solution section with geotextile.

Fig. 9. Application of geotextile.

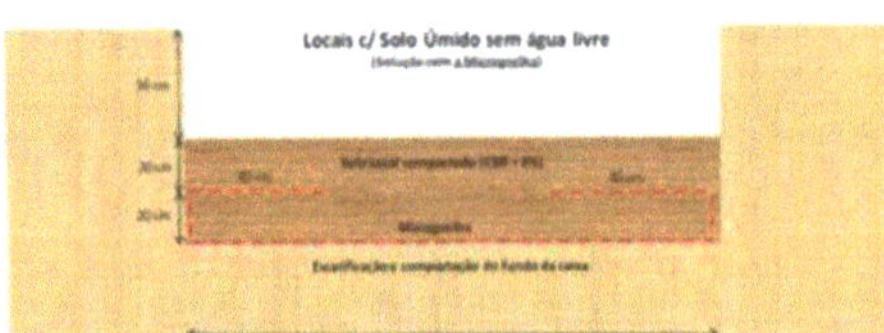

Fig. 10. Solution type section with microgrid.

Fig. 11. Application of the geogrid.

Figures 12 and 13 show the solution concerning the mixing of lime in the soil with the objective of improving soil characteristics compacted.

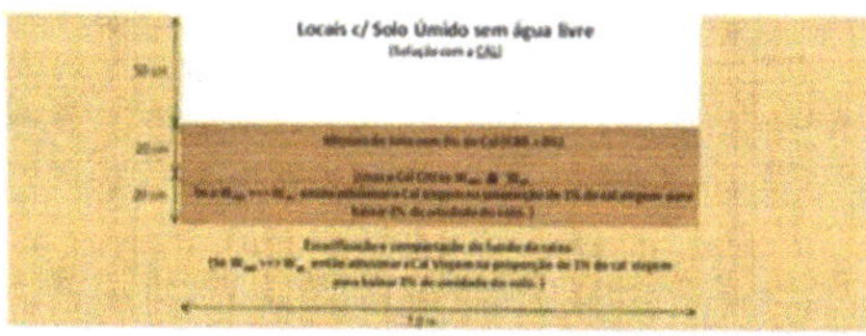

Fig. 12. Solution type section with lime.

6 Evaluation of Results

High deformability values were observed against some acceptable values in the solutions (Geotextile, Geogrid and lime). The Fig. 14 shows the execution of the test with Benkelman beam.

Therefore, an evaluation of the lower layers by means of the Penetrometer Dynamic Lightweight (DPL).

6.1 Evaluation of the Subleite with the DPL

The use of the dynamic penetration test with DPL aims to characterize and determine the resistance to penetration of soil deposits low capacity. This test helps to determine soil resistance in areas where the Palheta test becomes technically unfeasible, such as in granular soils, unsaturated soils and soils with high permeability.

At first, it was observed that soon below the surface layer (subleito) exists a soil that has low with high moisture content (Fig. 15).

Table 2 shows correlations between the N10 and soil consistency and indicates the presence of consistency from soft to medium in the study area.

Fig. 13. Application of lime.

Fig. 14. Benkelman beam test.

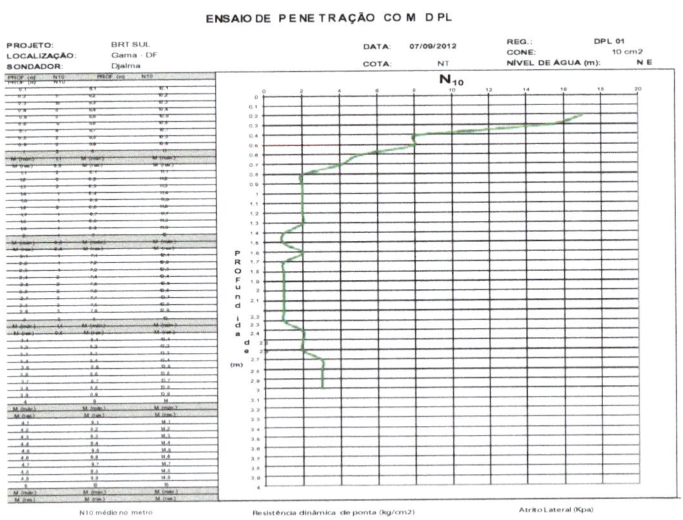

Fig. 15. DPL sondage of the evaluated area.

Table 2. Variation of soil resistance.

Vane Test	DPL	SPT	SPT	Consistência
Resistência (kPa)	N10	N30	N30	
Décourt 1989	DIN 4094*	Lambe/WhitmTerzhagi/Peck	Godoy 1972	–
<25	0–3	0–2	0–2	Muito mole
25–50	3–6	2–4	3–5	Mole
50–100	6–12	4–8	6–10	Média
100–200	12–22	8–15	11–15	Rija
200–400	22–45	15–30	15–19	Muito Rija
>400	>45	>30	>20	Dura

7 Conclusions

In general the solutions have to be feasible and bring the desired stability. Therefore the reinforcement of the deeper layers, the control of hair moisture, should be considered (Table 3).

Table 3. Summary of results.

Características	Camada estruturante (Rachão)	Reforço com Geotêxtil	Reforço com Geogrelha	Solo Cal	Cascalho
Redução da umidade de compactação	Não se aplica	Não se aplica	Não se aplica	Sim	Não se aplica
Atendimento as características de deformabilidade solicitadas.	Sim	Não avaliado	Não	Não	Sim
Homogeinezação da deformabilidade	Sim	Não avaliado	NÃO	NÃO	NÃO
Servir como camada drenante	Sim	Sim	NÃO	NÃO	NÃO
Redução da Capilaridade	Sim	Sim	NÃO	NÃO	NÃO
Velocidade de aplicação	Sim	NÃO	Sim	Sim	Sim

In view of the results presented, evident two solutions that can be applied:

a - Structural layer with allows for a need and therefore the increased thickness of the reinforced layer. The heterogeneity of layer deformability inferior is standardized by the "neg." against the service pressure used by the compaction equipment. Because according to with the variation of the surface resistance can penetrate more or less in accordance with Fig. 16.

Fig. 16. Variation of penetration (needling) as a function of soil surface resistance.

b- compacted gravel layer, seen that the deflectometric tests indicated good stabilization of the structure against the locations. This type of service depends on the humidity natural resources of the deposit, related to rainfall and transport distances, since the supply of this product is more scarce in the Federal District, compared to legal requirements for the release of Gravel. So that there is continuity earthmoving services, it is recommended that use of gravel during periods of drought.

References

Bernucci, L.L.B.: Considerações sobre o dimensionamento de pavimentos utilizando solos lateríticos para rodovias de baixo volume de tráfego. Tese (Doutorado) - Escola Politécnica, Universidade de São Paulo. São Paulo. 237 p. (1995)

Elliott, R.P., Thornton, S.I.: Simplification of subgrade resilient modulus testing. Transportation Research Record 1192 (1988)

Embrapa Cerados: Mapa pedológico digital—SIG atualizado do Distrito Federal escala 1.100.00 e uma síntese do texto explicativo, 31 p. (2004)

Medina, J.: Mecânica dos Pavimentos. Editora UFRJ, Rio de Janeiro/RJ, 380 p. (1997)

Motta, L.M.G.: Métodos de dimensionamento de pavimentos flexíveis; critério de confiabilidade e ensaios de carga repetidas. Tese de Doutorado. COPPE/UFRJ. Rio de Janeiro (1991)

Nogami, J.S., Villibor, D.F.: Uma Nova Classificação de Solos para Finalidades Rodoviárias. Simpósio Brasileiro de Solos Tropicais em Engenharia. COPPE/UFRJ, vol. 1, Rio de Janeiro, RJ, Brasil (1981)

Paranhos, H.S.: caracterização geotécnica dos principais grupos de solos do mapa pedológico do Distrito Federal (esc. 1:100.000): Estudo na área de dinamização urbana, entre Samambaia e Gama. Dissertação de Mestrado. UnB (1998)

Santos, J.D.G.: Contribuições ao estudo dos solos leteríticos granulares como camada de pavimentos. Tese (Doutorado) - COPPE/UFRJ. Rio de Janeiro, 342 p. (1998)

Author Index

H. Shehata and H. Poulos (Eds.): GeoMEast 2018, SUCI, pp. 269–270, 2019.
https://doi.org/10.1007/978-3-030-01917-4

FSC
www.fsc.org
MIX
Papier | Fördert gute Waldnutzung
FSC® C083411

Kolibri 360 GmbH
Ferdinand-Jühlke-Straße 7
99095 Erfurt, Deutschland
produktsicherheit@kolibri360.de